ECOLOGY AND EMPIRE

Ecology and Empire

Environmental History of Settler Societies

EDITED BY
TOM GRIFFITHS AND LIBBY ROBIN

UNIVERSITY OF WASHINGTON PRESS
SEATTLE

© in this edition, Keele University Press, 1997
Copyright in the individual contributions is retained by the authors

Keele University Press
22 George Square, Edinburgh

Published simultaneously in the United States of America by the
University of Washington Press, PO Box 50096,
Seattle, WA 98145-5096

Typeset in Linotype Stempel Garamond by
Carnegie Publishing, Preston and
printed and bound in Great Britain

A CIP record for this book is available
from the Library of Congress

ISBN 0 295 97667 5

The publication of this book was assisted by
Australians Studying Abroad and Qantas Airways Limited

The publications programme of the Sir Robert Menzies Centre
for Australian Studies is sponsored by BTR plc.

Acknowledgements

This book grew out of a co-operative project that was initiated while the editors were working at the Sir Robert Menzies Centre for Australian Studies in London. We thank the Menzies Centre, its former and current Heads, Professors Brian Matthews and Carl Bridge, and also the staff of the Institute of Commonwealth Studies for their support. Corporate sponsorship from Australians Studying Abroad and Qantas Airways Limited was greatly appreciated.

We are grateful to our publishers for their enthusiastic support for the project from its inception, especially Richard Purslow, Nicola Carr and Nicola Pike. We would also like to thank Rodney Mace, Edel Mahony, Luisa Pèrcopo and Alan Platt.

The greatest pleasure in editing this book has been in working with our contributors, all of whom have brought insight, energy and enthusiasm to the project.

Tom Griffiths and Libby Robin

Contents

Ecology and Empire: Towards an Australian history of the world

Tom Griffiths

'Ecology' and 'empire' are words that suggest very different dimensions of life on earth; at times they might appear to be opposites. One is natural, the other social; one is local and specific to place, the other is geographically ambitious; one is often seen to be scientific, amenable to laws and exclusive of humanity; the other is political, quixotic and historical. Brought together under the scrutiny of scholarship, these worlds and world-views make for creative friction. But 'ecology' and 'empire' also had a real relationship. They forged a historical partnership of great power – and one which, particularly in the last 500 years, radically changed human and natural history across the globe.

When, in 1986, the American historian Alfred W. Crosby wrote his important book, *Ecological Imperialism*, which built on his earlier *The Columbian Exchange* (1972), he threw those words together in his title and enjoyed the perversity of their pairing, the cheeky conjunction of apparent innocence and power.[1] His book described the *biological* expansion of Europe and saw humans as a species as well as political beings. That is also the aim of this collection, which acknowledges the inspiration of Crosby's work and is motivated by the conviction that environmental change has been, until recently, 'an unexplored aspect of colonialism'.[2] This book looks back at European expansion from the so-called colonized 'peripheries', the settler societies, and uses one of those societies – Australia – to shed new light on comparative environmental history. Crosby himself chose New Zealand as a case-study for his analysis of the intersections of ecology and empire; here we draw particularly on the histories of the United States, South Africa and Latin America, as well as Australia. This introductory chapter scrutinizes changing interpretations of ecological imperialism in the Australian setting to illustrate the way in which environmental histories of the 'edges' of empire are destabilizing traditional narratives of world history. The chapter then discusses concepts of 'settler societies', argues the value of comparing their environmental frontiers and histories, and, finally, introduces the five sections of the book.

Ecological imperialism and Australia

Alfred Crosby's book described how Europeans established themselves securely in far-flung but temperate countries and made them into 'neo-Europes': the United States, Canada, Australia, New Zealand, Argentina and Uruguay. These were the 'lands of demographic takeover', those countries where Europeans quickly became numerically dominant over the indigenous peoples, amounting to between 75 and nearly 100 per cent of the populations.[3] Why, asked Crosby, were Europeans able to establish such demographic dominance so quickly and so far from home? The answer, he argued, lay in the domesticated animals, pests, pathogens and weeds that the humans carried with them, an awesome accompaniment of colonizers that the settlers sometimes consciously nurtured and marshalled, but that often constituted an incidental and discounted dimension of imperialism. In Crosby's memorable words: European immigrants did not arrive in the New World alone, but were accompanied by 'a grunting, lowing, neighing, crowing, chirping, snarling, buzzing, self-replicating and world-altering avalanche'.[4]

Ecological Imperialism suggested that the superhuman achievements of European expansion were exactly that – more than human, and we have failed to realise just how much more. 'The human invaders and their descendants', wrote Crosby, 'have consulted their egos, rather than ecologists, for explanation of their triumphs.'[5] He depicted humanity as 'the purposeful but often drunken ringmaster of a three-ring circus of organisms'.[6] It is to the *passive* or *distracted* role of humans in ecosystems that he directed our attention, rather than to the manifest history of conscious social and political action that conventionally occupies historians.[7] Such an approach deliberately plays down the conscious and deliberate actions of humanity in order to reveal the independent and semi-independent dynamism of the natural world, itself normally the passive background in historical narratives. One danger of this approach is that, in extreme forms, it may present 'ecological imperialism' as a latter-day 'social Darwinism', a way of denying human agency – for good or ill – on the frontier. There is a genuine and important debate, particularly in the lands of demographic takeover, about the causes of 'the fatal impact', a debate often charged with emotion and politics. Historians argue about the number of deaths among indigenous peoples due to disease or violence, germs or guns, and wonder, too, whether even the introduction of disease were altogether accidental. Could smallpox, for instance, have been deliberately released amongst Australian Aborigines by early British colonists, a particularly sinister act of warfare?[8] Where did ecology end and imperialism begin? That question must not be avoided by the elision of the two words into a persuasive phrase. Ecological imperialism was sometimes a purposeful partnership and sometimes accidental; it was often both conscious and unconscious – here we recognize Crosby's image of the drunken ringmaster. In his book *Disease and Social Diversity*, Stephen J. Kunitz writes about the complex interplay of natural history and local history, of epidemiology and politics: 'There is no doubt that epidemic diseases had serious consequences ... but the most devastating contact situations seem to have been associated with dispossession from the land. Not all natives dropped dead when they got down wind of a European.'[9] Human responsibility remains a central and inescapable issue. But we are also reminded that humans are inextricably bound to the natural world, travel with more of it than they know, and often underestimate its independent historical influence.

Crosby himself played down the extent to which there was a two-way flow of life between the Old and New Worlds; the Australian eucalypt, for example, has been an impressive imperialist.[10] But he nevertheless asked a crucial question: why did the biotic conquests mainly favour the European? Why was 'the Columbian exchange' unequal? It is no mere academic question, for it was the stuff of everyday philosophizing and moralizing amongst settlers, who saw the evidence all about them. In Australia it generally led to an answer that was profoundly judgemental about Australian nature and peoples, and alternately confident and anxious about Australia's future. Ecology and empire, then, were not only factors shaping environmental realities on the frontier; they were also closely entwined in the settler psyche.

Australia revealed the paradox of the 'new lands' most sharply – it was newly discovered and settled by Europeans and it was new in the sense of being seen to be raw, unclaimed, unformed and full of promise. But the land, its nature and peoples, was also typecast as ancient, primitive and endemically resistant to progress. It was a land of living fossils, a continental museum where the past was made present in nature, a 'palaeontological penal colony'.[11] Marsupials like the kangaroo and koala and monotremes like the platypus and echidna were considered undeveloped or inefficient compared to placental mammals. Follow-ing 'cosmic laws', natives of all kinds were expected to 'fade away' in the face of exotics because they were inferior – and many settlers felt it wise to help such a process along in case the aboriginality of the country should reassert itself. Acclimatization societies systematically imported species that were regarded as useful, aesthetic or respectably wild to fill the perceived gaps in primitive Australian nature. This 'biological cringe' was remarkably persistent and even informed twentieth-century preservation movements, when people came to feel that the remnants of the relic fauna, flora and peoples, genetically unable to fend for themselves, should be 'saved'.[12]

Crosby's answer to the question of the success of migrating Europeans and imported biota demanded an adventurous excursion into the deep past. He reminded us that the underwater continental seams between the Old World and the New World had been crossed by humans in two distinct periods. The first crossings took place at various times during the 100,000 or more years before the ending of the last ice age, when humans discovered Australia from south-east Asia and others later crossed the Bering Strait into North and then South America. The second sustained period of crossings began just 500 years ago, when European voyagers discovered the unity of the sea and effectively turned the continents inside out.[13] In between these two crossings, explained Crosby, agricultural societies were established in various parts of the world, people lived together more densely, crops were systematically developed and harvested, and animals were domesticated. So, to quote Crosby: 'When Captain Cook and the Australians of Botany Bay looked at each other in the eighteenth century, they did so from opposite sides of the Neolithic Revolution.'[14] There are still echoes of nineteenth-century social Darwinism in his twentieth-century ecological imperialism: Crosby perpetuates the settler's perception of the Aborigines as a people 'without agriculture', and therefore somehow unfinished. But the agricultural revolutions of the Old World, he explained, had bred humans who were familiar with crowd diseases and a biota that was already disturbed, domestic and opportunistic, well adapted to colonizing and competing, and one that was bound to overwhelm a long-isolated ecology.

So there, on the eastern coast of Australia in the late eighteenth century, one of the great ecological – as well as cultural – encounters of all time took place. When the British arrived in New South Wales, their industrial revolution at home was beginning to gather pace, fuelled by the fruits of imperialism elsewhere. Therefore, Australia, unlike most other parts of the New World, experienced colonization and industrialization almost coincidentally, a compressed, double revolution.[15] Australia's later colonization made it the country to which George Perkins Marsh looked most hopefully for careful documentation of industrial society's environmental impact on virgin territory.[16] This was even more of a New World than America, which had once shared a land-bridge with Eurasia and still bore the marks of it. This was an encounter with a land that (unlike America) had never known hoofed, placental mammals, a land beyond what became known as Wallace's Line, an abrupt boundary of faunal types at Lombok, east of Bali and Borneo, that was identified in the mid-nineteenth century by the naturalist Alfred Russell Wallace. This was truly 'the antipodes', the newest continent but the oldest landscape, a late breakaway from Gondwana that had drifted for millions of years in a lonely evolutionary dance across the southern ocean. Crosby, an exponent of what, in this volume, John MacKenzie refers to as 'apocalyptic' environmental history, calls the encounter of 1492 'one of the major discontinuities in the course of life on this planet';[17] in the interests of competitive catastrophism, we might nominate 1770 (or perhaps we should say 1788) as an equivalently momentous date in world ecological history. The environmental historian George Seddon believes that the only comparable event to result in such biological instability was a major geological happening, the linking of North and South America in the Pliocene and Pleistocene, which enabled the invasion of South American marsupial fauna by placental mammals from the north.[18]

But it was for its triumphant social and political continuities that Australian history was first celebrated; it was written and presented as a relatively unproblematic footnote to empire. David Blair's *History of Australasia* (1878) argued that Australia offered a 'happy contrast' to the colonization of the

Americas and Africa, 'for no grander victory of Peace has this world ever witnessed than the acquisition of Australasia by the British nation'.[19] The continent's history began with British discovery in 1770, when 'a blank space on the map' – to quote the historian Ernest Scott – became tethered to the world.[20] In that year, the aimlessly drifting 'timeless land' was, for the first time, *anchored* – and by no less than the world's major maritime power. Australia had no 'history' of its own; only what was brought to it in ships.

The nineteenth century's emerging liberal vision of history, which depicted 'the sporadic but ineluctable advance of Freedom' from its natural home in Europe, gave the recently rescued peripheries a subordinate but significant place in world history.[21] British imperial and Commonwealth history provided a way of linking Australian history to other histories, a wider framework that persisted until the postwar era of decolonization, when it was dispersed and diminished into bookshop categories such as 'Asia', 'Pacific' and 'anthropology'.[22] In 1986, in an essay entitled 'The Isolation of Australian History', Donald Denoon, author of *Settler Capitalism*, drew attention to the loss of comparative frameworks for Australian history, producing what he called a 'partly self-imposed' isolation that impoverishes Australian history and also leads to a neglect of Australian scholarship by the rest of the world. This neglect, wrote Denoon, 'seems a terrible pity, since [Australian scholarship] offers arresting insights to non-Australians precisely *because* it is so difficult to locate in the context of conventional categories of experience'; he urged the development of historical approaches that would 'restore Australian experience to the rest of the world ... [and] reintegrate Australia into the history of humanity'.[23] One response to this sort of challenge has been to make those links either through Australia's deep indigenous past or its multicultural present. Having tired of 'dependency' as their umbilical cord to the world, Australians have recently discovered histories that are *pre*-colonial or *post*-colonial.[24]

From the very beginnings of the British occupation, colonists have questioned the depth and narrative potential of the indigenous past. As recently as 1969, however, Australian scientists and conservationists wrote a book called *The Last of Lands* – a phrase that encapsulates that paradox of the 'new lands', suggesting both 'recent' and 'primitive' – in which they described Australia as the last continent to be peopled and its nature, therefore, as unaffected by humans.[25] The scientific discovery of the antiquity of Aboriginal occupation of Australia now proceeded apace, and estimates quickly deepened from a few thousand years to tens of thousands of years. Australia, illuminated by the magic of radiocarbon, became what British archaeologist Grahame Clark called the most dramatic illustration anywhere of the physics of prehistory.[26] In the postwar world, reeling from the racial horrors of Hitler's war, human antiquity became a measure of human unity, a way of escaping from racial discourse and of locating a common, global past. 'To the peoples of the world generally', wrote Clark in 1943, 'the peoples who willy nilly must in future co-operate and build or fall out and destroy, I venture to suggest that Palaeolithic Man has more meaning than the Greeks.'[27] This was a different place for Australia in world history, one that would find commonality in its ancient past. But the growing articulation of post-colonial difference directly challenged this notion of universal culture and the authority of the West to reconstitute it. For Aboriginal people, the scientific search for antiquity was problematic for the very reason that it tethered ancient Australia to the world. It generalized a local story into a global one; it drew boundaries between the ancient past and the custodial present; it sketched historical, migratory connections between Aboriginal people and other humans, and ultimately challenged Aboriginal beliefs by finding Australia's human beginnings elsewhere.[28]

It is only recently that Australian history – once the whitest history in the world – has become cross-cultural in a most dramatic and revealing way. What was known as 'the Great Australian Silence'[29] has been broken, and stories of the violence and dispossession done to Aborigines have been allowed to be heard. The arrival of Europeans in Australia actually exploded a capsule of accelerating change; it initiated a process that was much less peaceful and more radical and oppositional than 'settlement', although that term itself had muted dimensions of conquest. Now 'settlement' has become re-envisaged as 'invasion', a shift in language and vision from outside to inside; the colonists become colonizers. In the words of Henry Reynolds, 'settled Australia ... is a landscape of revolution'.[30]

'Invasion' is a term that geographers used well before historians, because of their discipline's instinctive environmental orientation.[31] There is less of the historian's moral angst in their use of 'invasion', and more of the scientist's cool, biological perspective; the geographers knew that they were describing lurching frontiers of environmental practices and alien life-forms, and not just the calculated 'settlement' of political beings. If cross-cultural history has undermined the story of Australia as a footnote to empire, as a continental clock that started to tick in 1770,[32] then environmental history has further complicated the imperial narrative. As John MacKenzie argues in Chapter 14, the most recent phase of imperial natural historical writing has tended to see the era of European imperialism as but a brief period in the history of human interactions with tropical and subtropical ecologies. Such scholarship has revealed a much greater extent of environmental transformation by indigenous peoples than we had imagined, and it has discovered much longer cycles of environmental ups and downs with which the colonial moment has sometimes unknowingly interacted. These longer historical and environmental perspectives tend to diminish the apocalyptic power of the 1492s and 1770s and see them instead as but one of a series of encounters and transformations.

In the Australian context, I can think of no better examples of this school than Eric Rolls, Stephen Pyne and Timothy Flannery. Some people have puzzled at the extraordinary breadth of Eric Rolls's interests, in particular at the link between his two lifelong research topics, the history of Chinese in Australia and the environmental history of the continent.[33] But his chapter in this book helps to explain it, I believe, for in telling the story of China's centuries-old relationship with Australia, he breaks down our preconceptions of continental isolation in the distant past and the present, gives the 'multicultural present' a history, and dilutes our European fixation, our infatuation with 'remoteness' and 1770 and all that. It enables him to take a long view of Australian environmental history as a series of major 'disruptions'. The isolated and timeless land is found to have a history full of events and encounters. Rolls's vision builds on earlier work which includes what must surely be the best case-study of ecological imperialism in any of the settler societies, a book called *They All Ran Wild*.[34]

Stephen Pyne also takes a 250-million-year perspective on the fire history of Australia in a remarkable book called *Burning Bush*. 'However much Australians might lament their isolation', writes Pyne, '... there is no such quarantine for fire. Australian fire history is an indispensable chapter in a global epic that began when early hominids captured combustion and changed forever the human and natural history of the planet.'[35] Here, in Chapter 1, Pyne turns the torchlight back on the 'centre' and finds that Europe, which we are conditioned to see as the norm, is very much the exception in terms of its fire practices.

Timothy Flannery, too, generates an unusual view of Europe from the periphery. His book, *The Future Eaters*, takes a long-term view of the evolutionary predicament of the 'new' lands: New Holland, New Zealand, New Guinea and New Caledonia. Like Crosby, Flannery offers an invigorating history of the world and, in particular, an ecological account of human colonization across thousands of years. Flannery's distinctive parochialism allows him to go further, and to argue that the first arrival of humans in Australasia some 60,000 or more years ago – the crossing of Wallace's Line at Lombok – was 'an event of major importance for all humanity' and enabled a 'great leap forward' for our species as a whole.[36] Somewhere, about the time of the first colonization of Australia, humanity was transformed from being 'just one uncommon omnivorous species among a plethora of other large mammals' into the earth's dominant species. What was the cause? Flannery suggests that, by crossing Wallace's Line, humans discovered lands free of tigers and leopards and a biota unused to mammalian predators, where a managerial environmental mentality could blossom. Hence, according to Flannery, the first Australasians were the very first humans to escape 'the straitjacket of coevolution'; the consequent 'changes in technology and thought undergone by the Aborigines changed the course of evolution for humans everywhere'.[37] It is a startling and intriguing claim, an eminently debatable antipodean reversal.[38]

Flannery also builds on recent archaeological and anthropological work that has recognized the environmental interventions of Aboriginal people. When anthropologist Rhys Jones coined the term 'firestick farming' to describe Aboriginal land management, he deliberately and provocatively resuscitated that word 'farming' and applied it to a people allegedly 'without agriculture'. Flannery uses New Guinea to show that there, in contact with Australian

Aboriginal peoples, were some of the world's first agriculturalists 'at a time when agriculture was just a distant glimmer on the south-western horizon for Europeans'.[39] Different environmental pressures on the Australian continent led to a very different – and, to Europeans, an unrecognizable – type of farming. The extinction of Australia's megafauna may have been due to Aboriginal hunting or a result of habitat changes introduced by Aboriginal burning. Aboriginal culture, it emerges, was innovative as well as ancient; no longer can it be categorized simply as 'the stone age' of humanity, nor was it the quintessential hunter-gatherer society. In Australia were found the world's oldest cremation, perhaps the earliest human art, the first evidence of edge-ground axes, an early domesticated species in the dingo, millstones that predated agricultural revolutions elsewhere, and the most ancient evidence of modern humans.[40] Flannery's zoologist's eye enables him to perceive humans as a species and to generalize both Aborigines and Europeans as 'future-eaters', both exploiters and managers of nature. In the words of George Seddon: 'the most important fact in the environmental history of Australia is that it had a radically new technology imposed upon it, suddenly, twice.'[41]

Flannery's book, and his chapter in this collection, provide a detailed Australasian extrapolation of Crosby's thesis. He confronts us with truths about our land that we have not yet fully assimilated: that Australia has the poorest soils in the world, a stressful, unreliable climate, a fragile and heavily interdependent ecology, and great biodiversity. He explains more completely than Crosby why introduced species overwhelmed Australian natives, and he does so by again reversing one of our cultural stereotypes, by depicting Europe not as 'home' or 'the centre', but as 'the Backwater Country'. 'If we are to understand Australasian history properly', argues Flannery, 'we must understand a little of the ecology of the Europeans.'[42] So the Australian gaze turns back across the world and, ecologically speaking, sees a comparatively raw and rapacious biota. It is Europe that is actually the 'new land', more recently colonized by *Homo sapiens* than Australia, with a simplified biota that had to start again after the last ice age and is now populated by invasive, dominating weeds, animals and plants that were pre-adapted to disturbed environments. Now we know that they were weeds before they even left, not just when they spilled out of the ships on to Australia's ancient soils!

Crosby's depiction of Captain Cook and the Australians as regarding one another from opposite sides of a sharply defined Neolithic revolution is already looking too simple. The radiating ripples of the imperial model are already eddying back from the 'edges' and muddying the waters.

Settler frontiers and environmental history

By referring in my title to 'an Australian history of the world', I am not seeking to replace one form of imperialism with another, but drawing attention to the way in which Australians, more conscious now of their indigenous natural and human history and of its depth and integrity, have won back some agency in the global narrative. Furthermore, the Australian experience makes the interactions of ecology and empire a central historical problem and demands a more complex (and distinctive) account of Crosby's ecological imperialism. Scholars, including the contributors to this book, have begun to use the insights of local ecology and history to fragment and overturn the conventional patterns of imperial history. There is much to be gained from an awareness of the parallels and differences between settler societies and the creative dialogues at the edge of empire. The more we learn from these, the less these places look like edges. Richard Grove's *Green Imperialism* is a wonderful stimulus to this perspective. It traces the ways in which the environmental experiences of the colonial periphery were not purely destructive, but generated pioneering conservationist practices and new European evaluations of nature.[43] In Chapter 9, Grove continues this questioning of 'the more monolithic theories of ecological imperialism'.[44]

In planning this volume and in using the concept of 'settler society', we have incorporated many of the 'lands of demographic takeover' and deliberately reached beyond them. The inclusion of perspectives from South Africa (part of the Old World – and of the New), Mexico and the Caribbean signals our determination to learn also from those settler societies where European immigrants never became numerically dominant yet were able to gain a disproportionate amount of power, maintain a viable political constituency, and assert and defend their strength through explicitly racial institutions. In these societies, the racial politics that was more easily marginalized or denied in the 'lands of demographic takeover' became clearly visible, even inescapable, and ecological control by the settler state became more complex.[45] In South Africa, Jane Carruthers's demonstration of the socially divisive potential of that ecological 'good', the national park, and Shaun Milton's description of settler-owned cattle as 'shock-troops' give an Australian (or North American) historian a jolt, first of surprise and then of recognition.[46]

'Settler society' conveys political investment in a land that goes beyond mere commercial expansion of an empire. The Australian geographer Archibald Grenfell Price (1892–1977), whose *oeuvre* anticipated Crosby's interest in 'moving frontiers of diseases, animals and plants', distinguished between 'sojourner' and 'settler' colonization.[47] The American geographer Donald W. Meinig, building on Price's work, described the difference between 'settler empires' and 'commercial sea empires'. A settler empire, in Meinig's words, was the 'permanent rooting of Europeans in conquered soil' – an investment, a 'plantation', that meant that 'settler colonies took on a life of their own to a degree quite unparalleled by any other type of imperial holding'.[48] Australian use of the term 'settler' suggests that it carried valued transformative meanings, connotations of 'progressive frontiering'. It persisted as a word in rural districts, often in place of 'farmer', beyond pioneering days, well into the twentieth century. Settled land was contrasted with 'the back country', wild land or land where one might find Aborigines 'at large'. The 'settler's clock' was the kookaburra,[49] and the strips of bark that hung from eucalypts were the 'settler's matches'. 'Settling' could also be contrasted with 'squatting' because it suggested the establishment of families and community; it brought social institutions and political bureaucracy. This meaning of 'settlement' was curiously extended to describe Aboriginal communities administered by a public authority: 'Settlement blacks' implied Aborigines who had been enveloped and disempowered by bureaucracy.[50] A settler society, whether or not numerically dominant, was an invading, investing, transforming society with an internal frontier, both natural and cultural.

Efforts to understand and reinvigorate the concept of 'frontier' have been fundamental to the recent emergence of environmental history as a sub-discipline, particularly in the United States. Historians such as Donald Worster, William Cronon, Richard White and Patricia Limerick have set out to reveal the environmental dimensions of conquest and to rediscover nature as an active agent in the making of the American West.[51] Their work is exciting and richly suggestive, but sometimes surprisingly nationalistic. Our aim is not to rival their nationalism and seek to replace the American frontier with the Australian one, but to build on their stimulating work by using a different frontier as a prism through which to filter the writing of environmental history. A fine recent example of such an enterprise is William Beinart and Peter Coates's *Environment and History: The Taming of Nature in the USA and South Africa*. Frontier history is one of the few well-developed areas of comparative scholarship, and Australian, American and South African frontiers in particular have often been studied in parallel, though rarely from an environmental perspective.[52]

Australia's frontier was called 'the outback', 'the Inland', 'the back country', 'the outside track', 'our backyard', 'the Never-Never', 'the Dead Heart' or 'the Red Centre': the descriptive metaphors are about hearts and backs, but never about heads or fronts. The Australian frontier could be heroic and colourful and character-forming like America's West; it could be an object of natural or spiritual pilgrimage like America's West; it could generate distinctive national stereotypes like America's West; it had equivalents of cowboys, Indians and log cabins. But the westering in Australian history was not nearly as sustained or progressive as America's, the Great Plains notwithstanding; settlement ebbed and flowed and regularly confronted its limits; the Australian frontier could never be said to have 'closed' as America's was said to be in 1890. The Never-Never never ended: the American dream was the Australian nightmare.[53] 'We seemed to be looking round the bend of the earth', wrote the scientist Francis Ratcliffe of the Australian saltbush country in 1935; 'later I was to be really scared – scared that something in my mind would crack, that the last shreds of my self-control would snap and leave me raving mad.'[54] The Australian pastoral frontier, argues Flannery, was an artefact of megafaunal extinction, a grassland resource that, for tens of thousands of years, could not be fully exploited by humans in the absence of large herbivores, and which provided the British settlers and their flocks with a short-lived bounty, an ecological niche that was exhausted in their lifetimes.[55] Australian settlers have often been forced to retreat or have had to be enticed out there by the government; the antipodean parables of the plains are far less about the entrepreneurial freedom celebrated in America, and much more about national and racial anxiety. The open spaces, people said, needed to be talked up; they needed to be developed and populated for the defence of nation and the defence of race.[56] 'Populate or perish!' was the familiar cry, and those who challenged this maxim or publicized Australian aridity, such as the geographer Griffith Taylor, were accused of being unpatriotic. J. M. Powell has done more than any other Australian scholar to explore this aspect of the Australian geographical imagination[57] and, in Chapter 10, Brigid Hains describes some of these anxieties at work on two very different Australian frontiers – the ice of Antarctica and 'the Inland' of Australia.

Thus, for Australian settlers, 'ecology' and 'empire' represented the competing realities of geography and history, land and culture, and stood for a fundamental, persistent tension between origins and environment in Australian life. 'Australia is antipodean, not Australasian', claimed one geographer in 1963, meaning that Australia was really Europe down under and remained free of what he called Aboriginal or Asian 'adulteration'.[58] The most famous contortion created by this tension was the 'White Australia' policy, a defensive statement about the biotic future of the country and an official government stance against the region in which Australia found itself.[59] Grenfell Price's work on 'moving frontiers' and the 'changing landscapes of greatly improved cultural and other characters' that followed in their wake was informed by his defence of this policy.[60] In the second half of this century, however, Australians almost accidentally, and then with growing conviction, became a 'multicultural' nation; in the same years, paradoxically, they became more critical of their 'multinatural' inheritance.[61] Today, Australians discuss whether they can and should become 'part of Asia'. Yet Australia is as different ecologically from Asia as it is from Europe. What, then, is the future cultural and political significance of that great ecological border, Wallace's Line?

'Ecology' in this volume does not just mean the portmanteau biota of the Europeans and the distinctly local environments that they encountered; ecology is also the lens through which we claim to be reinterpreting imperial history. The science of ecology itself was partly an artefact of empire, as Libby Robin and Thomas Dunlap show. And ecology is the science that environmental historians have used energetically as their metaphor and model, a relationship that distinguishes the newer environmental history from the older and sometimes neglected traditions of historical geography. This intriguing disciplinary disjunction, which this book aims to help bridge, may have been partly due – as Michael Williams recently suggested – to a temporary abandonment of 'environment' by many geographers in the 1950s, 1960s and 1970s, just before historians awakened to the promise of the field.[62] Describing a 'new teaching frontier' in 1972, the American historian Roderick Nash felt 'that the environmental historian, like the ecologist, would think in terms of wholes, of communities, of interrelationships, and of balances'.[63] But ecology and environmental history have helped to shape one another, and it has been partly the historical studies of natural communities that have since undermined these models of community and balance and of perceived distinctions between stable nature and disturbing humanity.[64]

Environmental history often makes the best sense on a regional or global scale, rarely on a national one. The major environmental forces that have shaped Australia, for example, come into focus on analytical levels other than that of the nation – by seeing Australia as a settler society, as part of the New World frontier, or as a continental cluster of bioregions. Environmental histories of Australia, therefore – and of other countries, too – need continually to fragment or enlarge the national perspective and to scrutinize and reflect upon the intersections of nature and nation. The term 'settler society' signals our focus on the emigrant European cultures and biota and their interactions with indigenous peoples and nature, rather than on indigenous perspectives and experiences themselves.[65] Some of the most interesting current work in imperial environmental history, however, draws on indigenous environmental knowledge and politics, past and present, and fruitful collaborations are developing between western scientists and local indigenous peoples.[66] Some of that work informs this book, but its focus remains on settler societies, ideologies and ecologies. Michael Williams reminds us that imperialism is neither solely western nor only international, and Richard Grove breaks down the category of the settler 'us'.

This book is divided into five parts. The first, 'The Ecologies of Invasion', begins with a fire history of the expansion of Europe and then describes the ecology and invasions that have shaped Australia. The work in this section illustrates two new perspectives that have enlivened the environmental history of settler societies: a long-term view of evolving ecosystems and environmental transformations at the periphery; and a historical and ecological appraisal of Europe itself, seeing the comparative novelty of some of its norms. It also introduces the Australian environmental experience, to which later chapters refer. Part 2, 'The Empire of Science', scrutinizes ecology itself, but also sees 'science' as necessarily inclusive of technology, a partnership apparent in settler societies, where science in *place*, rather than abstract, universal science, is a driving, creative force. Correspondences and tensions between local and imported science (in the broadest sense) are the concern of the contributors to this section, who are keen to dilute and complicate the historical relationship between 'centre' and 'periphery' by drawing attention to the substantial lateral intellectual and technological exchanges that occurred between settler societies themselves. Part 3, 'Nature and Nation', addresses some of the popular and political ecological visions to be found in settler societies, in particular racial and ethnic definitions of nature. Notions of 'empire' and 'nation' are unpacked to reveal nature as a political and psychological tool and versions of ecological imperialism that are Afrikaner, Scottish and Australian, rather than amorphously European. Part 4, 'Economy and Ecology', brings together a variety of empires and a range of ecologies to show how invading market economies confronted and interacted with pre-existing and enduring indigenous forms of industry and commerce. The economy did not just disrupt ecology. Bankers could not ignore biota (nor the biota, bankers). A global system of exchange and production was not simply imported and imposed. There were racist markets, tenacious organic economies and contradictions between imperialist ideology and local ecology and economics. The final section of the book, 'Comparing Settler Societies', provides a broad historiographical perspective on imperial environmental history and the place of this book within it, reviewing themes and issues that percolate throughout its chapters.

If the imperial framework is naturally comparative, the colonial one is instinctively nationalistic; each can be illuminated by the other perspective. The heart of empire needs to be reminded of its own embedded settler histories and environmental frontiers; the 'new lands' need to be more conscious of one another and of the historical and philosophical correspondences between them. 'Ecology and empire went hand in hand', writes a group of American historians of the West, 'and we are only just starting to understand their relationship.'[67] This book puts that relationship under the microscope and pursues its wayward offspring around the globe.

Notes

1. Alfred W. Crosby, *Ecological Imperialism: The Biological Expansion of Europe, 900–1900* (Cambridge: 1986), and *The Columbian Exchange: Biological and Cultural Consequences of 1492* (Westport, Connecticut: 1972).
2. William Beinart and Peter Coates, *Environment and History: The Taming of Nature in the USA and South Africa* (London: 1995), p. 2.
3. Alfred W. Crosby, *Germs, Seeds and Animals: Studies in Ecological History* (New York: 1994), p. 29.
4. Crosby, *Ecological Imperialism*, p. 194.
5. Crosby, *Germs, Seeds and Animals*, p. 41.
6. Crosby, *Germs, Seeds and Animals*, p. xiii.
7. Alfred W. Crosby, 'The Past and Present of Environmental History', *American Historical Review*, 100/4 (1995), p. 1177.
8. Noel Butlin, *Our Original Aggression* (Sydney: 1983); Alan Frost, *Botany Bay Mirages* (Melbourne: 1994), chapter ten.
9. Stephen J. Kunitz, *Disease and Social Diversity: The European Impact on the Health of Non-Europeans* (New York: 1994), p. 178.
10. See MacKenzie, Chapter 14, below.
11. Quoted in Adrian Desmond, *Archetypes and Ancestors: Palaeontology in Victorian London, 1850–1875* (Chicago: 1982), p. 104.
12. 'Biological cringe' is an adaptation of a famous Australian phrase, 'cultural cringe', and was recently used by Nick Drayson in 'Comparing Australian Animals: Australia and the

"Biological Cringe"', paper presented to the British Australian Studies Association Conference, Stirling, 30 August 1996.

13. J. Parry, *The Discovery of the Sea* (Berkeley: 1981); Michael Williams, 'European Expansion and Land Cover Transformation', in Ian Douglas, Richard Huggett and Mike Robinson (eds), *Companion Encyclopaedia of Geography* (London: 1996), pp. 182–205.

14. Crosby, *Ecological Imperialism*, p. 18.

15. For Latin America's contrasting, sequential experience of these phenomena, see Elinor G. K. Melville, Chapter 12, below.

16. George Perkins Marsh, *Man and Nature*, edited by David Lowenthal (Cambridge, Mass.: 1965), p. 49; David Lowenthal, *George Perkins Marsh: Versatile Vermonter* (New York: 1958); see Michael Williams, Chapter 11, below.

17. Crosby, 'Reassessing 1492', in *Germs, Seeds and Animals*, p. 185.

18. George Seddon, 'The Man-Modified Environment', in John McLaren (ed.), *A Nation Apart* (Melbourne: 1983), p. 16.

19. Quoted in Henry Reynolds, 'Violence, the Aboriginals and the Australian Historian', *Meanjin*, 31 (1972), p. 471.

20. Ernest Scott, *A Short History of Australia*, 4th edn (Melbourne: 1920), p. v.

21. William McNeill, 'The Changing Shape of World History', in Philip Pomper, Richard H. Elphick, and Richard T. Vann (eds), *World Historians and their Critics*, special issue of *History and Theory*, 34 (1995), p. 11.

22. Donald Denoon, 'The Isolation of Australian History', *Historical Studies*, 22/87 (1986), p. 252; survey of London bookshops, 1995–6.

23. Denoon, 'The Isolation of Australian History', pp. 254, 260.

24. But see J. B. Hirst, 'Keeping Colonial History Colonial: The Hartz Thesis Revisited', *Historical Studies*, 21/82 (1984), pp. 85–104.

25. L. J. Webb, D. Whitelock and J. Le Gay Brereton, *The Last of Lands: Conservation in Australia* (Brisbane: 1969).

26. Grahame Clark, *Space, Time and Man: A Prehistorian's View* (Cambridge: 1994), p. 124.

27. Grahame Clark, 'Education and the Study of Man' (1943), in his *Economic Prehistory* (Cambridge: 1989), pp. 410, 416.

28. Tom Griffiths, 'In Search of Australian Antiquity', in Tim Bonyhady and Tom Griffiths (eds), *Prehistory to Politics: John Mulvaney, the Humanities and the Public Intellectual* (Melbourne: 1996), pp. 42–62.

29. W. H. Stanner, *After the Dreaming* (Sydney: 1969), p. 13.

30. Henry Reynolds, *Frontier: Aborigines, Settlers and Land* (Sydney: 1987), pp. 192–3.

31. For example, see Andrew Hill Clark, *The Invasion of New Zealand by People, Plants and Animals* (New Brunswick: 1949); A. Grenfell Price, *White Settlers and Native Peoples* (Melbourne and Cambridge: 1950), and *The Western Invasions of the Pacific and its Continents* (Oxford: 1963).

32. Michael Cathcart, 'The Geography of Silence', *RePublica*, 3 (1995), p. 180.

33. For Eric Rolls's work on Chinese in Australia, see *Sojourners* (St Lucia, Queensland: 1992), and *Citizens* (St Lucia: 1996).

34. Eric C. Rolls, *They All Ran Wild* (Sydney: 1969).

35. Stephen J. Pyne, *Burning Bush: A Fire History of Australia* (New York: 1991), p. xiii.

36. Timothy Fritjof Flannery, *The Future Eaters* (Sydney: 1994), p. 153.

37. Flannery, *Future Eaters*, p. 14.

38. For some of the debate, see an outstanding review of the book by George Seddon, 'Strangers in a Strange Land', *Meanjin*, 2 (1995), pp. 290–301, and a reply by Lesley Head, 'Meganesian Barbecue', *Meanjin*, 4 (1995), pp. 702–9.

39. Flannery, *Future Eaters*, p. 298.

40. Tom Griffiths, *Hunters and Collectors: The Antiquarian Imagination in Australia* (Melbourne and Cambridge: 1996), p. 93.

41. Seddon, 'Man-Modified Environment', p. 10.

42. Flannery, *Future Eaters*, p. 303.

43. Richard H. Grove, *Green Imperialism: Colonial Expansion, Tropical Island Edens and the Origins of Environmentalism, 1600–1860* (Cambridge: 1995).

44. Grove, *Green Imperialism*, p. 7.

45. William Beinart, 'Introduction: The Politics of Colonial Conservation', *Journal of Southern African Studies*, 15/2 (1989), p. 147; see Shaun Milton, Chapter 13, below; Tom Johnson, 'Colonial Ecology, Agricultural Policy and the Structure of Knowledge in 1930s Northern Rhodesia', paper presented to Human Ecology Seminar, Department of Anthropology, University College London, 7 October 1996.

46. Chapters 8 and 13, below.

47. Price, *The Western Invasions*, chapter 3.

48. D.W. Meinig, 'A Macrogeography of Western Imperialism: Some Morphologies of Moving Frontiers of Political Control', in Fay Gale and Graham H. Lawton (eds), *Settlement and Encounter: Geographical Studies Presented to Sir Grenfell Price* (Melbourne: 1969), pp. 213–40. See also Donald Denoon, *Settler Capitalism* (Oxford: 1983), and 'Settler Capitalism Unsettled', *New Zealand Journal of History*, 29/2 (1995), pp. 129–41; and Nathan Reingold and Marc Rothenberg (eds), *Scientific Colonialism* (Washington, DC: 1987), pp. vii–x.

49. Graeme Davison, *The Unforgiving Minute: How Australia Learned to Tell the Time* (Melbourne: 1993), p. 28.

50. W. S. Ramson (ed.), *Australian National Dictionary* (Oxford: 1988).

51. William Cronon, *Nature's Metropolis: Chicago and the Great West* (New York: 1991); Patricia Nelson Limerick, *The Legacy of Conquest: The Unbroken Past of the American West* (New York: 1987); Richard White, *'It's Your Misfortune and None of My Own': A New History of the American West* (Norman and London: 1991); Donald Worster, *Dust Bowl: The Southern Plains in the 1930s* (New York and Oxford: 1979), *Under Western Skies: Nature and History in the American West* (New York and Oxford: 1992), and *Rivers of Empire: Water, Aridity, and the Growth of the American West* (New York: 1985).

52. For example, H. C. Allen, *Bush and Backwoods: A Comparison of the Frontier in Australia and the United States* (East Lansing, Mich.: 1959); Fred Alexander, *Moving Frontiers* (Melbourne: 1947); Robin W. Winks, *The Myth of the American Frontier: Its Relevance to America, Canada and Australia* (Leicester: 1971); Howard Lamar and Leonard Thompson (eds), *The Frontier as History: North America and Southern Africa Compared* (New Haven: 1981).

53. I am grateful to David Lowenthal and Michael Williams for suggesting these contrasts.

54. Francis Ratcliffe, *Flying Fox and Drifting Sand* (1938; Sydney: 1947), p. 260. For more on Ratcliffe, see Chapters 4 and 5, below.

55. Flannery, *Future Eaters*, pp. 392–3.

56. For South African parallels, see Shaun Milton, Chapter 13, below.

57. J. M. Powell, *An Historical Geography of Modern Australia* (Cambridge: 1988), and *Griffith Taylor and 'Australia Unlimited'* (St Lucia: 1993). See also Chapter 7, below.

58. J. Macdonald Holmes, *Australia's Open North: A Study of Northern Australia Bearing on the Urgency of the Times* (Sydney: 1963), p. xi.

59. On 'ecological independence and immigration restriction', see Philip J. Pauly, 'The Beauty and Menace of the Japanese Cherry Trees: Conflicting Visions of American Ecological Independence', *Isis*, 87 (1996), pp. 51–73.

60. For an assessment of Price's writing and career, see J. M. Powell, 'Archibald Grenfell Price (1892–1977)', in T.W. Freeman (ed.), *Geographers: Bibliographical Studies*, 6 (1982), pp. 87–92.

61. On Australian 'ecological purity', see David Lowenthal, Chapter 15, below.

62. Michael Williams, 'Environmental History and Historical Geography', *Journal of Historical Geography*, 20/1 (1994), pp. 3–21, esp. p. 9; see also J. M. Powell, 'Historical Geography and Environmental History: An Australian Interface', *Journal of Historical Geography*, 22/3 (1996), pp. 253–73.

63. Roderick Nash, 'American Environmental History: A New Teaching Frontier', *Pacific Historical Review*, 41 (1972), pp. 362–71.

64. See the articles by Donald Worster, Richard White and William Cronon, in 'A Round Table: Environmental History', *Journal of American History*, 76/4 (1990).

65. But see Melville, Chapter 12, below.

66. For example, Farieda Khan, 'Rewriting South Africa's Conservation History: The Role of the Native Farmers Association', *Journal of Southern African Studies*, 20/4 (1994), pp. 499–516; Peter Latz, *Bushfires and Bushtucker: Aboriginal Plant Use in Central Australia* (Alice Springs: 1995); J. R. W. Reid, J. A. Kerle and S. R. Morton (eds), *Uluru Fauna* (Canberra: 1993).

67. William Cronon, George Miles and Jay Gitlin, 'Becoming West: Toward a New Meaning for Western History', in *Under an Open Sky: Rethinking America's Western Past* (New York: 1992), p. 12.

PART I

Ecologies of Invasion

CHAPTER I

Frontiers of fire

Stephen J. Pyne

Introduction: the labours of Linnaeus

In 1749 Carl von Linné was at the height of his fame – professorially ensconced at Uppsala, popularly enshrined as a 'second Adam' who had named all the plants and animals of the world, celebrated for his travels throughout Sweden, especially Lapland, renowned as the author of *Systema Naturae* and other classics of natural history written under his Latin *nom de plume*, Linnaeus. So it was natural that King Frederick I should invite him to tour Sweden's most southerly lands, Skåne, that Linnaeus should agree, and that he should use the occasion to return through Småland, his home province.

There he studied the contentious subject of *svedjebruk*, Swedish 'swidden', 'looked upon by some as profitable, by others as rather deleterious'. So prevalent was the practice that some Linnaean contemporaries believed that Sweden had derived its name from the endless *svedje* that comprised its countryside. Pondering the evidence, Linnaeus concluded that, while fire unquestionably consumed humus ('the food of all growth'), the practice allowed farmers to get 'an abundance of grain ['and for several years ... a good pasture of grass'] from otherwise quite worthless land'. Deny them that burning, and 'they would want for bread and be left with an empty stomach looking at a sterile waste ... a thankless soil and stony Arabia infelix'.

The passage, however, outraged Baron Hårleman, High Commissioner of Agriculture, and Linnaeus was forced to delete it from the final publication. 'Not only', Hårleman fumed, had Linnaeus 'not condemned *svedjebruk*, so pernicious for the country, but even contrary to his own better judgment justified and sanctioned the undertaking'. In penance, Linnaeus was compelled to insert a long passage on the value of livestock manure as a way of supplementing traditional forest composts of heather, moss, and conifer needles. Even in High Enlightenment Europe, it seems that burning could not compete with bullshit.[1]

As a dialectic between humans and nature, fire regimes express the values, institutions, and beliefs of their sustaining societies. The labours of Linnaeus tell us a great deal about European fire, about the extent to which it was embedded in a social matrix of agriculture, and, hence, why it looks invisible to those, like North Americans and Australians, to whom fire is recognizable only when it free-burns in wildlands and bush. But the episode also alerts us to the peculiar problems that would arise when a civilization like Europe's encountered lands far more fire-prone than its sodden core and dealt with folks far more fire-addicted than European ministers. Compared with Aboriginal burning in Australia, agricultural and pastoral fires in Africa and India, landscape-sweeping burns across North America, and colonizing fires in new settlements every-where, *svedjebruk* was as quaint as a midsummer's day bonfire. So, too, the outrage provoked by rural burning in tiny Skåne would pale before the dialogue inspired by continental-scale firing that confronted an imperial Europe.

Those differences matter. Fire on earth looks the way it does today because Europe expanded beyond its constricting peninsulas and islands to become a global power and because it vectored to that overseas imperium an industrial revolution that could, by exploiting fossil hydrocarbons, transcend the endless ecological cycles of its agricultural heritage. Europe has influenced fire in Asia, Africa, Australasia, and the Americas in ways in which none of those places has influenced European fire. From Europe's ancient association of fire and agriculture came conceptions of fire's proper place in the landscape. From Europe's alliance of forestry and imperialism came the attempt to suppress fire in large forest and wildlife reserves created in overseas colonies. From Europe's industrialization came the apparatus to enforce the agenda of fire abolitionists. It is no accident that the continued condemnation of fire by international environmentalism – from nuclear winter to greenhouse summer, from fire as an emblem of social disorder to fire as a perverter of biodiversity – has its origins in Europe. Europe's fire has become, as Europe always believed it would be, a vestal fire for the planet.

An endless cycle: fire and fallow

Before there was a Europe, there was fire. But as the last of the Pleistocene glaciations receded, that fire was anthropogenic and, increasingly, it was pres-cribed by the imperatives of the Neolithic revolution. The hominid colonization of Holocene Europe was a flame-catalysed reclamation by agriculture. Over long millennia, livestock and cultigens, tended by fire-wielding humans, pene-trated into every valley, propagated up every slope, prevailed over every alternative biota. The social order dictated the biotic order. Europe became an immense garden.

There was no lack of fire. In mediterranean and boreal Europe, agriculture only replaced one fire regime with another. In central Europe – the temperate core of the continent – the Neolithic vigorously expanded the dominion of fire. Here agricultural colonizers shattered the biotic bell-jar of shade-lording lindens and elms and concocted a biotic stew that simmered over a succession of landscape-cooking fires. Fire had helped to establish the garden; fire cleaned it periodically; fire powered the dynamos of its nutrient cycles. Without fire, much of Europe was uninhabitable. Practitioners never doubted this fact.

Europe was a dense mosaic of landscapes and peoples, and the geography of its fire reflected this complexity. Yet almost all Europe's agricultural systems, excepting (with qualifications) irrigated croplands, were variants of fire-fallow farming. Some lands lay fallow for a season, some for one year out of a handful, some for decades while cropland succumbed to rough pasture and then to woods. There is no good name for this collectivity of fire-rekindled fallowing. *Swidden* is the common anthropological expression, but when it was introduced to the literature to describe east Asian practices in the 1950s, it was greeted with disdain (and dismay). Critics like H. H. Bartlett preferred indigenous terms, of which there were dozens. Why, he asked, accept for the Philippines and the East Indies an antiquated expression drawn from Viking-colonized Yorkshire to describe the burning of the ling? In 1958 the word remained so obscure that it was not even recorded in the *Oxford English Dictionary*. But the term triumphed, anyway, a metaphor for the ascendancy of European concepts throughout the globe. It seems only appropriate to have it apply, with all its faults, to Europe itself.[2]

Europe's fallows, like its poor, were always there. Intellectuals hated them. With few exceptions – Vergil, Linnaeus, a handful of others who had grown up on farms – professional agronomists, all trained in cities and housed in academies and bureaux, detested the fallows as a waste of productive land and an invitation to sloth. Worse, the fallows were burned. For a civilization constantly pushed to (and beyond) its demographic limits, haunted by visions of famine and hunger-driven disease, the flaming fallows were unconscionable. The 'disgrace of the fallows,' as François de Neufchâteau declared in the eighteenth century, obsessed agricultural intelligentsia and officials. Why did they persist? Why could they not be utilized instead of being sacrificed to the faithless flames? Why could not cultivation proceed to the point that it dispensed with fire altogether?[3]

There was no single reason. But agronomists had the relationship reversed. Fire did not follow fallow, as plague did rats. The burning was not a convenient (if, to official eyes, indolent and reckless) means of disposing of agricultural trash. The fallows were not burned because they had overgrown; they were grown in order to be burned. They were cultivated to support fire, just as three-field rotations grew oats and barley to feed draft animals. Fire was integral, not incidental. Agriculturalists relied on plants to recapture nutrients and then burned them to liberate those biochemicals in a suitable form. They needed to jolt fields back to life, needed to purge soils temporarily of hostile micro-organisms and weeds, needed to flush stale pasture with succulent proteins. Fire alone did this and, in order to burn, fire required suitable fuels; these were grown. Even constantly cultivated infields required outfields from which to gather combustibles or to run the herds from which, housed in winter barns, manure could accumulate. Those outfields were swiddened and their rough pastures burned.

Fire and fallow constituted an endless cycle, now swelling outwards, now contracting, but never broken, which informed European agriculture and thereby shaped the European landscape and informed a European land ethic. Where officials saw environmental and social wastage – lost revenues, wandering swiddeners and pastoralists, incinerated soils, scorched timber – peasants saw renewal; their fire ceremonies clearly spoke to the dual virtues of fire to purge the bad and promote the good. Fire was as essential to the farm as to the household, a tool more indispensable than ploughs and harrows. By contrast, an urban intelligentsia experienced fire in cities and identified it with social disorder, especially war. Intellectuals denounced; peasants burned.

The official protest did have its logic. However much agriculture reshaped landscapes, it could not, ultimately, make something out of nothing. The prevailing understanding of nature was that it was profoundly cyclic. Growth and decay were exquisitely balanced; since the Creation, no species had been added to or subtracted from the great chain of being; agriculture required that no more could be removed from a field than had been put into it. Productivity could only increase by building up the reservoir of soil nutrients. Failure to return as much as was harvested led to degradation, of which there were endless examples. In this Newtonian ecology, landscapes orbited with a biotic balance as delicate as that traced by the planets between gravitational and centrifugal forces.

Each group perceived the central fire differently. What farmers saw as a biotic hearth, agronomists saw as an ecological *auto-da-fé* which burned away the carpets and walls. Agricultural fire condemned civilization to a rural rut and prevented any hope of improvement. That fire burned away humus was, therefore, sufficient cause to criticize it; that it promoted wasteland and fallow was enough to condemn it. Fire took away: it did not give back. The transmuting fire was, to its critics, a kind of folk alchemy, proposing to turn environmental lead into agricultural gold. Traditional fire practices seemed no better than superstition, rituals with no more substance than the ancient fire ceremonies that burned witches and heretics. For millennia, fire traced the great divide between 'rational' and 'primitive' agriculture.

Nature's economy was inseparable from society's. It did not help fire's cause that within a landscape like Europe's, shaped by human artifice, wildfires appeared most prominently during times of social breakdown. War, revolution, famine, pestilence – anything that left the garden untended – would let fallow run riot and would end, like peasant rebellions, in torch and sword. Society and land were both bound by the same inexorable logic; and fire threatened each of them. Even the lordly Linnaeus could not escape that cycle, nor did he wish to. The same year that he toured Skåne, he published his widely influential essay *Oeconomia Naturae* (The Economy of Nature), in which he conveyed the exquisite checks and balances that informed nature's polity and that equally governed cultivation. The theme obsessed him. In his last, brooding essay, he pondered the character of divine justice by which every deed had, with almost Newtonian logic, its retribution.[4]

But there was no doubting fire's presence and power. When Linnaeus was still a youth, Herman Boerhaave had declared that 'if you make a mistake in your exposition of the Nature of Fire, your error will spread to all branches of physics, and this is because, in all natural production, Fire ... is always the chief agent.' Around that central fire Europe's agricultural systems orbited. Even long hunters in Finnmark, transhumant pastoralists in the Apennines and Pyrenees, and Slavic swiddeners in Siberia were caught in its ecological force field like swarms of biotic comets.[5]

Extending the cycle: enlightenment and expansion

But if Linnaeus himself could not escape, his apostles – the twelve students he dispatched around the world – could. Symbolically, they extended the circle of the European Enlightenment and, with it, a dynamic of exploration and empire that established a European hegemony not only in the world's political economy, but in its scholarship. That propagating periphery was all too often a frontier of fire.

Europe reconstructed the way in which it thought about the world and began to rebuild the world accordingly. Intellectual Europe increasingly accepted modern science as the model of knowledge; it enshrined progress, not renaissance, as nature's informing principle. Imperial Europe renewed rivalries on a global scale in what William Goetzmann has termed a 'Second Great Age of Discovery'; overseas outposts moved beyond trading factories on the coast and probed boldly inland, repopulating landscapes with European émigrés, remaking foreign lands into a constellation of neo-Europes, a colonial outfield to the metropolitan infield. Industrial Europe, building on a century of agricultural reformation, experimented with an economy no longer circumscribed by humus, manure, and sunlight.[6]

Carrying the torch: new lands, new fires

The Maori, the Malagasy, the Madeirans – all have founding myths of a great fire that accompanied settlement, and the record of such fires is buried in their soils and lakes. Similar contact fires have left comparable records in Brazil, Iceland, Australia, everywhere that had colonizing peoples and the means to preserve charcoal. That imperial Europe should also have its world-transmuting fires is no surprise. Everywhere the strike of European steel on indigenous flint threw sparks to all sides. Forested frontiers, in particular, were a flaming front of eruptive fires with names like Peshtigo and Miramichi and Black Thursday which left behind a landscape of more subdued, residual combustion.

And fire *did* remain, sometimes a cause of these immense changes, often a consequence, always a catalyst. Europe's expansion brought it to lands notoriously susceptible to fire – to environments for which seasonality meant an oscillation between wet and dry, not cold and hot; to biotas salted with pyrophytes; to landscapes already baked in a hominid hearth. In places, colonization expanded fire by transferring European fire practices to receptive landscapes, or by breaking down old biotas and adapting native burning, reconstituting those lands with Eurasian surrogates. So swidden exploded out of New Sweden across the American backwoods frontier; sheep swirled around the gyre of Australian grasslands; livestock crushed the once-grassy deserts and savannas of the American West and the South African Karoo. The exotic fauna soon spread exotic flora, from wild artichokes to cheat grass to tumbleweeds, and fires rose and fell with the strength of these biotic tides. Elsewhere, colonization sought to contain or even extinguish indigenous fires.

Ultimately, settlement *exchanged* one fire regime for another. Colonizers could not, in the end, deny the logic of an earth fluffed with combustibles and marinated in oxygen. There might be more fire or less fire, there might be efforts to suppress certain kinds of undesirable fires kindled by natives, malcontents, and ignorant immigrants, but there was no expectation that fire itself would disappear, or that fire practices could vanish like old smoke. Fire would simply be glossed into the text of the new landscape, the rubrics of a flame-illuminated manuscript.

Certainly, this was true for North America and Australasia. The redefinition of fire regimes is exactly what C. S. Sargent's map of American forest fires for the 1880 census, Franklin Hough's 1882 *Report on Forestry*, and hundreds of settler journals all corroborate. America, Canada, and Australasia were developing countries, most of whose peoples lived by agriculture. Conflagrations might rage during the time of transition, but great wildfires required wildlands; they would cease as settlement became sedentary and as controlled burning found its niche in the cultivated countryside. Meanwhile, there was little doubt that fire was an inextricable presence and that its proper use was the best means of fire control.

Fire conservancy: protecting land, preventing fire

Still, Europe's colonies presented some novel circumstances which magnified Europe's ancient fire dialectics. There was a natural desire to control the worst fires rather than to rely on their *laissez-faire* recession; and there was general recognition that the new worlds offered opportunities for wholesale reform on a scale never before imagined. There were lands too inimical to sedentary farming – their climates too dry and unstable, their soils too impoverished, their weeds and pests too potent. Such lands could not be farmed easily and, in the case of mountains (so experience suggested), should not be.

In such instances, colonizing powers began to reserve lands, usually forested, for the public good. These were an imperial invention; they were possible because the lands were (or could partially be made to be) vacant; and they created an imperative for institutions to administer them. Especially in Asian and African colonies, fire joined famine, malaria, and banditry as a perceived blight on lands mired in fatalism and rural inertia, for which railroads, telegraphs, civil service bureaus, and fire protection could promise hope, progress, and humus-laden soils. What the British called 'fire conservancy' had analogues in all those places to which the imperium of the European Enlightenment spread.

Had those landscapes simply disappeared into a reconstituted agricultural regime, the earth's combustion calculus would bear little resemblance to today's. The geography of free-burning fire – by which I mean fire that responds to conditions of terrain, weather, and vegetation – among the industrial nations traces precisely the creation of official wildlands. It is worth noting how extraordinary these institutions are, how exceptional their presence in world history, how fragile their survival. The oldest do not date beyond the 1860s. Most have a tenure of less than a century. New Zealand has already begun their disestablishment. Reversions, some large, are promised to indigenous peoples in Canada and Australia.

The wholesale reservation of lands for the public good was an obvious, if idealistic, solution to human misuse. Remove humans and you remove abusive human practices. (There was little sense that much of what was attractive about colonial landscapes had resulted from long human manipulation.) The particular arguments for park reserves were aesthetic and nationalistic and, for forest reserves, to regulate logging and ameliorate the climate; both sought to decouple the ecological syllogisms by which colonization seemingly led to catastrophe, and deforestation to degradation. On previously uninhabited islands from Madeira to Mauritius, Europeans had immediately introduced fire and, as the vegetable mould went up in smoke, so, it seemed, the local climate had become droughty, springs had dried up, rare flora and fauna had perished, and once-Edenic isles had degraded into cinders. Those microcosmic experiments foreshadowed macrocosmic doom as similar scenarios appeared to be unfolding before the eyes of European savants wherever colonization, like a steel wedge, had cracked open the continents. The most direct way in which to intervene was to deny access to pioneering peoples.[7]

What made such reservations politically plausible was that the land was more or less emptied of humans. Those landscapes that were largely vacated following European settlement – in North America, Australasia and, in a different sense, Siberia – became the particular landscapes for public reserves. Disease had mostly done the task; where it failed, relocations, colonial and internecine wars, and general social disruptions had pruned populations and prevented a demographic recovery. Because of this population collapse, however, the anomalous had become the norm for a period of time: people had gone or were rapidly going. In Australia and in particular North America, the great era of land reservation occurred precisely at this instant, when the indigenes were fading and the colonizing Europeans had not yet arrived in great numbers. A few decades either way and it is unlikely that those reserves could have been made. They were a magnificent historical accident.

One important consequence was that, to exploring Enlightenment *philosophes*, that landscape (notably in North America) seemed preternaturally wild. Relic bands of indigenes appeared no more competent to shape the scene than did those minor streams which occupied the great valleys previously scoured by Pleistocene glaciers. The land appeared fresh from the Creation. Natives seemed to dissolve into it like salt in the ocean. In reality, this perception was a freak of historical timing. The reality was that the land had been as fully occupied as technologies had allowed, that most places were intensively shaped by indigenous practices, that many landscapes in the Americas and Australia were as fully anthropogenic as any found in Europe, that much of the New World had experienced human settlement for far longer than had the Old. But, during contact, the land had gone feral. Colonizing Europeans had, as one critic expressed it, discovered a garden and left behind a wilderness.

Where the indigenous peoples persisted, the reserves were always compromised. Even a handful of graziers or nomads could wield enormous power. Here, reserves endured because of political force, backed by legal statute or military cantonment. Such quasi-inhabited landscapes continued to burn. This was as true of southern crackers in Florida and smallholders in Gippsland as of Ghond tribesmen in central India. Whatever the legal regulations regarding access, the reserves were biologically inaccessible without suitable fire; the locals burned. Moreover, fires could ecologically void edicts, fences, and patrols. Once the natives recognized the value that officials placed on fire control, they had a ready weapon of protest. Arson of woods became endemic.

Where the reserves remained essentially uninhabited, however, there was no obvious problem with anthropogenic fire. Nor was there an imperative to instal a new system of controlled burning. Agricultural fire was necessary only if one lived by farming, grazing, hunting, or foraging on the land. The administrators of the great reserves did not: they only protected those landscapes. Accordingly, they were led to an anomaly as great as the vacant lands themselves. They could see the exclusion of fire as not only admirable, but possible.

Forestry and fire: the paradox of fire protection

That foresters, not *jägmeisters* or agronomists or civil engineers, inherited the task of administration had enormous repercussions for the reserves. Colonial foresters saw fire through a peculiar prism and almost always found themselves fiercely at odds with indigenous practices, all of which exploited fire. It is no accident that the first question to be asked at the first symposium on the first large-scale experiment in forest reserves – the 1871 forest conference in India – was the necessity and practicality of fire control. 'There is no possible doubt', wrote Lieutenant-Colonel G. F. Pearson, Conservator of Forests for the Central Provinces, that 'the prevention of these forest fires is the very essence and root of all measures of forest conservancy.' That sentiment was echoed by foresters everywhere.[8]

Paradoxically, this was not the case in Europe itself, whose woods continued to simmer over chronic fires. Silviculture, after all, was a graft on the great rootstock of European agriculture, and modern forestry was part and parcel of the agricultural revolution that had preceded the better-known industrial. Agro-forestry was the norm, not a novelty; the fire practices of agrarian Europe were also typical of its woods. Even in 1870, as much as 70 per cent of the Schwarzwald was subject to a swidden cycle that involved cereals, root crops, and oak, for which fire was fundamental. At the end of the nineteenth century, French foresters around Provence regularly practised *petit feu*, in which strips were protectively burned in regular rotation. Controlled burning persisted in the Ardennes forests until World War I, in the woodlands of the Midi until the 1920s, and in the Baltic pineries, rich with heath, until World War II. Fire was as prevalent – and as essential – in Europe's forests as in its fields.

But if woodsworkers exploited controlled fire, forest overseers hated and feared it. The political dialectics that informed agriculture – the divide between theorists and practitioners, field officers and government ministers, periphery and metropole – also shaped forestry. The brief against fire was that it destroyed trees better suited for timber or charcoal than for ash, that it encouraged graziers and swiddeners to encroach on protected woods, that it eliminated the humus which was the universal index of ecological health. In nature's economy, fire was a prodigal heir who spent his nutrient capital rather than living off his annual interest. Following fire, biotas degraded; soils eroded; weeds proliferated; hillsides slid; torrents rushed; climates degenerated; and societies became un-stable. The glory that was Greece, the grandeur that was Rome, the mystery that was the Maya – all succumbed to the internal rot of deforestation by that unholy alliance of cutting, grazing, and burning. To Europe's labourers, fire was a good servant but a bad master. To Europe's rulers, fire made for rich parents but poor children. Yet for centuries the only practical means of fire control remained close cultivation and prescriptive use.

This was no less true in nineteenth-century settler societies. The best hope was to change fire regimes as painlessly as possible, not to eliminate fire altogether. Study Sargent's 1880 fire map of America and you see a developing nation, profoundly agricultural, remaking itself with steam and industrial capital, but one for which any talk of fire exclusion was utopian nonsense. The character of fire had changed and would change further. While no one argued that conflagrations like those that swept Gilded-Age Wisconsin and Michigan were desirable, or wanted to repeat the holocaust that seared Victoria in 1851, no sane critic argued that fire itself could be expunged. When, in 1898, Gifford Pinchot thundered that 'the question of forest fires, like the question of slavery, can be postponed, at enormous cost in the end, but sooner or later it must be faced', no one took his abolitionist analogy literally, not even in Gippsland, then suffering through the conflagration that became known as 'Red Tuesday'. The problem was to subdue wildfire, and the solution, as always, was to tend the garden and burn the fallow carefully.[9]

Counterfire: last fires, lost fires

Yet imperial Europe often pursued a radically different fire ambition in its colonies than in its own cultural hearth. Ideals long confined to hypothetical islands, and practices safely caged in ancient social contexts were transported across the seas and released, propagating like the rabbits on Porto Santo or incongruously imposed like utopian colonies planted in Paraguay. Why? One explanation is political. European powers were prepared to behave differently towards colonial peoples than towards their own. European forces could control indigenes in Cochin, Natal, the Maghreb, and New South Wales in ways in which they could not control peasants in Galicia, Dalarna, and Provence. By the mid-nineteenth century, forestry had, in fact, become an inextricable part of European imperialism.

Colonial forestry was a composite, a kind of intellectual and institutional plywood that glued together the separate veneers of Germanic silviculture, French *dirigisme*, and British imperialism. The Germans were supreme as silviculturalists; the French had welded forestry to the purpose (and power) of the state; and the British – who at the start of the nineteenth century had lacked both forests and foresters – had fused the two into a package suitable for export. For Greater India, it hired German forest conservators, trained students at the French school at Nancy, and shipped British cadets throughout the empire. (When that imperium collapsed, so, almost overnight, did British forestry.) Until then, they promulgated a transnational culture of fire control, carried equally to Cyprus, Sierra Leone, Tasmania, the Cape Colony, New Zealand, even Hong Kong. It is instructive to recall that the founders of American forestry, like Gifford Pinchot, Henry Graves, and Theodore Woolsey, Jr., passed through this same curriculum; that Pinchot corresponded at length with Sir Dietrich Brandis, doyen of the Indian Imperial Forestry Service (and honoured as the 'Father of American Forestry'), about how to establish such an organization; that Pinchot later remarked that American foresters in the Far West had much in common with French colonial foresters in Algeria(!).[10]

That unfortunate comment should rattle our consciousness and remind us of extent to which the public lands and public forestry were an imperial invention. Even in Europe foresters existed primarily as state officers, either to promote wholesale reclamation of wastelands like the Landes, the Hautes-Alpes, or the Jutland heaths, or to intervene between local economies and the larger commercial forces against which traditional practices had too often proved incompetent. After the French Revolution had abolished the old forest regulations, an orgy of cutting and burning had prompted an environmental Reign of Terror that had left many communal forests in ruins and, more ominously, had stripped alpine slopes to the point at which torrents propagated like swarming locusts. State-sponsored reforestation became the preferred remedy further consolidating the confederation between forestry and the nation-state. The alliance between forestry and the state, planted in Europe, thrived overseas. Prominent foresters behaved like proconsuls, moved in the highest circles of imperial administration, and were knighted for their service to the empire. Forestry acted as an enlightened despot for the environment. Much of the American West, for example, was first settled by foresters through the institutions of fire protection.

It is instructive that the major imperial powers of Enlightenment Europe were from Europe's temperate core, from lands that did not experience routine fire seasons, that understood fire as an artefact of agriculture. The great powers were also the dominant industrial nations and scientific authorities and they often dismissed indigenous knowledge as shamanism, and folk practices as ecological acupuncture. For colonial landscapes that had known only traditional practices, European forestry was a revolutionary force. For colonial authorities, it was a means of wholesale biotic rationalization and social reform. Through forestry, the colonizers would transform irregular wildlands into regularized woodlands, just as the institutions of European jurisprudence would transform jungles of folk mores and as the railroads would restructure subsistence economies. Fire control was as fundamental to colonial rule as military garrisons, plantations, and acclimatization societies.

This, however, is still an argument to seize the torch, not to extinguish it. It does not explain why the foresters believed that they should seek to abolish fire. One reason is that they could only envision fire as anthropogenic. Everywhere they looked, human burning had overwhelmed and defined the landscape; controlling fires and controlling people were one and the same task. Where reserves had successfully excluded humans, however, it seemed possible that fires, formerly tolerated as a necessary evil, could likewise be banished. Only later did the potential for lightning fires become problematic. (In a sense, by removing the domesticated fire, foresters would allow the feral fire to replace it.) Fire exclusion seemed plausible because of the exclusion of people. The exemplar of uninhabited isles like tropical Mauritius proved less an ideal than an aberration, a *fata morgana* of agricultural *philosophes*, and so, too, fire control untrammelled by normal politics – call it imperial Europe's suppression paradigm – shimmered over colonial forests like blue haze.

A second reason is that the foresters themselves did not live in and off those reserves. They protected land; they did not cultivate it. Had they been forced to inhabit those landscapes, they would have been compelled to manipulate them with fire, as people everywhere did – as, in fact, their professional brethren in central Europe did. They would have had no choice. But they were guards, not gardeners. So it seemed possible – given sufficient political will – to do in the colonies what was only quixotic in Europe, to remove fire entirely from the garden. In those colonies densely populated with indigenes, the scheme failed; in those lands effectively depopulated, the experiment could proceed for several decades before its ecological and economic costs became overwhelming. Having banned the domesticated fire, officials allowed the feral fire to replace it.

Thus, forestry behaved differently outside its originating lands. Like so many other European utopias, forestry's was necessarily situated across the Western

Ocean. The suppression paradigm, released from its originating social ecology, spread like blackberries in New Zealand or cheat grass in the Great Basin. In Europe proper, the quarrel over fire practices was ancient, and the balance of land-use power prevented a massive extinction of burning. What finally snuffed out those flames was the spread of industrialism, the fossil fallow of coal. The Schwarzwald swidden, for example, expired when steam transport rendered its oak-derived tannic acid no longer competitive against South American imports.

In Europe's colonies, the confrontation over fire practices was starker and shorn of traditional checks and balances. Clearly, some form of fire protection was mandatory, but the form it should take was not obvious. The collision of European forestry with indigenous landscapes sparked a public debate about appropriate fire practices and policies. The celebrated light-burning controversy in America had cognates in the early-burning debates that were kindled throughout the British Empire. All ended with official condemnation of burning; with the political clout to attempt to enforce that edict even beyond the reserves themselves; and, after 40–60 years, with a recantation.

The reasons for that universal failure are all too familiar. The world's biotas obeyed different rhythms from those of temperate Europe; especially in those places where landscapes experienced strong gradients dividing wet and dry seasons, fire persisted in defiance of agronomic and political theory, and ecosystems displayed fundamental adaptations to it. Fire exclusion rendered biotas less stable, less useful, less diverse, and less amenable to fire control. For a while, fire protection could be made to work. For a period of time, installing a first-order infrastructure for fire prevention – by eliminating indigenous fires, by suppressing long-smouldering fires, and by actual firefighting – could dramatically reduce burned acreage. But that period of grace would not last. Either the land was converted to some other, less flammable, condition or else some species of controlled burning had to be introduced.

The mistake everywhere was not that Europeans sought to impose a new fire regime or that they fought wildfires to that end, but that they sought fire's abolition. They failed to recognize that fire's removal was as powerful an ecological act as its introduction. They believed that fire's suppression would liberate oppressed biotas, much as the suppression of famine, typhoid, and Thuggee could free backward societies to progress. Over and over again, they interpreted fire in political rather than environmental terms, as the graffiti of ecological vandals, as the torches of barn-burners and rural vigilantes, as the protest of a folk both sullen and prescientific. The tragedy in America, in particular, was not that wildfires were suppressed, but that controlled fires were no longer set.

Even so, early fire control necessarily relied on fire use. Until some alternative pyrotechnology appeared, fire exclusion was a concept as metaphysical as the geography of Gog and Magog, or Thomistic arguments for the existence of angels. Until then, paradoxically, the imperial Europe that had sought to abolish free-burning fire had created, with its immense colonial reserves, the ideal circumstances for fire's perpetuation.

Transcending the cycle: industrial combustion and fossil fallow

When Carl von Linné died in 1778, the Linnaean landscape of fallow and field, with its ecology of closed cycles and unbroken great chain of being, was beginning a rapid disintegration. Two years earlier, Captain James Cook had sailed on his last voyage, confirming an immense age of European exploration and empire; Thomas Jefferson had written the Declaration of Independence, not only inaugurating an era of democratic revolution, but announcing a colonial break-out from the coast into the interior, a vanguard of neo-Europes populated by emigrant settlers; Adam Smith had published *The Wealth of Nations*, proposing a new economic order; and James Watt and Matthew Boulton had consolidated their partnership to manufacture steam engines. The closed loops of the Linnaean world fractured; expansion metamorphosed into evolution; nature's cycles became spirals, and society's orbits the time-lines of progress.

Of these momentous events, the steam engine was, for fire history, the most profound. Its demand for fuel soon exhausted biomass reserves and compelled the custodians of industrial combustion to exhume lithic biomass from the geologic past. First coal, then petroleum and gas, were a kind of biotic bullion that acted on nature's economy like the plundered wealth of the Aztecs and Incas did on imperial Spain's. No longer was combustion limited by the self-regulating ecology of grown fuels. Fossil fallows could replace living ones; steam transport could restructure the flow of nutrients so that ecosystems aligned with the routes of commerce and the trophic flow of capital; novel pathways of energy could be built on the controlled combustion of fossil fuels. Industrial pyrotechnologies could replace the fire practices of open burning. It seemed possible to transcend the closing circles of earthly ecology. It was even possible, apparently, to apply the new combustion not only in exchange for, but to suppress, the old.

This was critical. However fanatical foresters might be regarding free-burning fire, they could only change it, not destroy it, if they wanted to retain forest wildlands. Fire *protection* meant redesigning fire regimes, not abolishing them; fire *control* required burned fuelbreaks, pre-emptive burning around protected sites, and regular *petit feu*; fire*fighting* meant burning out and backfires. In the end, fire remained. In India, imperial foresters discovered to their dismay that, in order to secure prime groves, native guards were early-burning all the surrounding forests. Fire exclusion was possible, at best, for only a few years. European forestry could no more escape fire than could European agriculture; forestry was a long-fallow swidden, growing oak, pine, and sal instead of wheat, turnips, and cotton. Eventually, fire would return. Eventually, that is, until industrial combustion arrived.[11]

With power pumps, fire engines, aircraft, tractors, power saws, and motorized transport, it became possible to move firefighters into the reserved lands quickly enough to catch fires while they were small and to meet free-burning fire, at its early stages, with a counter-fire force of equal magnitude. It was possible to impose a relative condition of fire exclusion for a longer duration. This cost money, it demanded ever larger investments of technology and firefighters, and it wrenched the biota into successively greater distortions; but it could be done. In the short term – and with state sponsorship for which costs were not matched against the values protected – suppression by rapid detection and inital attack was the most effective means of fire protection. The extent of areas burned from wildland fire plummeted. For settler societies, America replaced Europe as the oracle and exemplar of fire protection.

In fact, fire has receded everywhere in the developed world, telegenic conflagrations notwithstanding. Fires have disappeared from domestic life, from industrial pyrotechnologies, from agriculture, from forestry, and from many wildlands. Wherever fire is a *tool*, it has had to compete with the new pyrotechnologies and has usually lost. Whatever its ecological merits, open burning exists within an anthropogenic landscape for which free-burning fire and flame's obnoxious side-effects (like smoke) are no longer desired or even tolerated. The free fall of free-burning fire promises to stabilize only where fire is essential as an ecological *process* for which no surrogates are possible. Fire has retreated, like grizzly bears and Karner blue butterflies, into special preserves of public or Crown land. The geography of fire has become one of massive maldistribution – too much wildfire, too little controlled burning; too much combustion, too little fire. By the 1990s, America's public lands were immersed in a crisis of 'forest health' that was provoked, in good measure, by a fire famine.

Of course, the ideology of fire control mattered. Without the vision of fire exclusion, there would be in Europe's settler societies today more anthropogenic fire, less ecological havoc, and a better equilibrium between fire use and fire control. But the larger trajectory of fire history argues that industrial pyrotechnologies were destined to substitute wherever possible and that free-burning fire would recede into those habitats deliberately reserved from human habitation. The most profound of the modern epoch of extinctions may be that of anthropogenic fire. The epic of settler society fire, like all epics, is a tragedy.

Conclusion: vestal fires and virgin lands

Linnaeus could never have imagined a world without fire. On his travels, he met wildfire in Lapland, and throughout cultivated Sweden he encountered controlled fires for farming, forestry, grazing, tar, potash, lime, charcoal, every imaginable human endeavour. No less than its woods, the age structure of Sweden's towns also betrayed their fire history. As the hearth shaped the house, so the dynamics of house fires had shaped cities. Some landscapes needed more fire, some less, and others a different regimen. But fire there would be. The abolition of fire was as hypothetical as the extinction of water or dirt.

Europe's authorities fretted over fire not only because it threatened property (and lives), but because it challenged a social order whose environmental expression was the garden. Too often, fire encouraged movement. Swiddeners, pastoralists, long hunters, trappers, settlers – all exploited fire as part of seasonal or secular travels which shredded principles of fixed property ownership and mocked a social order in which, as with garden crops, everyone had his time and place. Free-burning fire seemed an index of social disorder, feeding on the scraps dropped from the high table of sedentary agriculture and on the weeds growing through the cracks of crumbling or ill-tended civilizations.

But that mobility, that capacity to render new, was precisely the moral premise behind settler societies. For them, flame epitomized the transmutational process; the torch was a transitional device handed from indigene to colonist; the fired landscape was the cultural hearth for a new people. The character of the land was important to settler societies. Their nationhood stories told of encountering, transplanting, and remaking the discovered lands into something similar to the Europe they had left and, in the end, something better.

Europe wanted the vestal fire of the hearth relocated, not rekindled. But rekindling is exactly what the émigré colonies required and what they did, and, accordingly, fire entered into the moral universe of settler societies, as agency, as index, as synecdoche, as symbol. Even as the settlers dutifully carried the flame from the metropole's hearth, they had to feed it new fuels and, in the process, they altered the character of the defining fire itself. From their virgin lands they nurtured a new vestal flame. Yet there, as elsewhere, fire followed fuel.

Their national identities derived in particular from the existence of vacated, 'new' landscapes. Its wilderness made America distinctive from Europe; its ineffable bush rendered Australians something more than Europeans in exile; its veld assured African colonists that they could never be subsumed under a strict European order. But settler societies could not have those wildlands without having fire also. Their free-burning fires – magnificent, reckless, sublime, devouring – were an immense burnt offering which commemorated the experiences of their complex origin; the fact that they were so different from Europe's domesticated, garden-variety burns made them all the more valued as archetypes. The Yellowstone fires of 1988, the Ash Wednesday fires of 1983 – these were cultural epiphanies, as fully an expression of national values as the stock-market crash of 1929 or the Burke and Wills expedition of 1860–1. Europe had no cognate, could have no equivalence, could hardly understand, much less appreciate, the wild magnificence and horror of the spectacle.

Eventually, settler societies realized that their landscapes were themselves anthropogenic, that both first-contact scenes and those hammered by hard labour into colonial forms were, to a large extent, the outcome of human hand and mind. Even wilderness had emerged from the forge of anthropogenic fire, not from the pristine spark of nature. But the differences with Europe remained, regardless; undeniable, ineffable, defining. Wildlands and fire could not be segregated: to confront one was to confront the other. Selecting suitable fires was another way in which cultures defined who they were. For settler societies, those historic fires of contact were as much a part of their cultural heritage as battles, exploring naturalists, the national novel, or exotic flora. Their perpetuated fires were a brilliant if, at times, macabre vestal flame. And because, as Patrick White wrote in *The Tree of Man*, 'They had looked into the fire, and seen what you do see, they could rearrange their lives. So they felt.' [12]

So they still feel.

Notes

This derives from a manuscript being prepared for publication by the University of Washington Press. Its full title is *Vestal Fire: An Environmental History, Told through Fire, of Europe and of Europe's Encounter with the World*. I have restricted my citations here to direct quotations. The book, when published, will contain the full complement of ancillary annotation.

1. The episode is well documented in Swedish sources. An excellent summary (with English translations) is available in Gunhild Weimarck, *Ulfshult: Investigations Concerning the Use of Soil and Forest in Ulfshult, Parish of Örkened, During the Last 250 Years*, Acta Universitatis Lundensis, Sectio II, 1968, no. 6 (Lund: 1968), pp. 45–7.

2. See Harold C. Conklin, 'An Ethnological Approach to Shifting Agriculture', *Transactions of the New York Academy of Sciences*, 17 (1954), pp. 133–42; H. H. Bartlett, *Fire in Relation to Primitive Agriculture and Grazing: Annotated Bibliography*, vol. 2 (Ann Arbor: 1957), p. 511.

3. Quoted in Marc Bloch, *French Rural History*, trans. Janet Sondheimer (Berkeley: 1966), p. 213.

4. The full title is *Specimen Academicum de Oeconomia Naturae submittit I. J. Biberg* (Uppsala: 1749). The standard English translation is available in Benjamin Stillingfleet, *Miscellaneous Tracts Relating to Natural History, Husbandry, and Physick*, 2nd edn (London: 1762), pp. 37–129.

5. Boerhaave is quoted in Gaston Bachelard, *The Psychoanalysis of Fire*, trans. Alan C. M. Ross (Boston: 1964), p. 70.

6. See William H. Goetzmann, *New Lands, New Men: The United States and the Second Great Age of Discovery* (New York: 1986), and, for a somewhat different interpretation, Stephen Pyne, 'Space: The Third Great Age of Discovery', in Martin J. Collins and Sylvia K. Kraemer (eds), *Space: Discovery and Exploration* (Washington, DC.: 1993), pp. 14–65. For Linnaeus's apostles, see Wilfred Blunt, *The Compleat Naturalist* (New York: 1971), pp. 183–91.

7. The best treatment of these island 'experiments' is Richard H. Grove, *Green Imperialism* (Cambridge: 1995).

8. Lt. Col. G. F. Pearson, *Report on the Administration of the Forest Department in the Several Provinces under the Government of India, 1871–1872* (Calcutta: 1872), p. 9.

9. Gifford Pinchot, 'Study of Forest Fires and Wood Protection in Southern New Jersey', *Annual Report of Geological Survey of New Jersey* (Trenton, NJ: 1898), Appendix, p. 11.

10. See Gifford Pinchot, *Breaking New Ground* (Seattle: 1972), especially pp. 6–9; Herbert Hesmer, *Leben und Werk von Dietrich Brandis 1824–1907* (Opland: 1975), pp. 327–84; Pinchot's observations are in Theodore S. Woolsey, Jr., *Studies in French Forestry* (New York: 1920), p. vi.

11. An interesting corroboration of this fact comes from a Russian fire specialist who spent six weeks in California in 1994 and offered as his first impression the observation that Americans relied on machinery more than Russians, who typically resort to counter-firing: 'American and Russian methods of fighting large forest fires are profoundly different. Because the Russian Federal Forest Service does not have a sufficient number of airtankers and helicopters for direct attack methods, control of an initial attack fire is frequently achieved using a backfire.' Remove that power equipment (and the money behind it), and you would see fire restored, although not in a form that burning advocates would prefer. Quotation from Alexander K. Selin, 'Forty-two Days in California', *Wildfire* 5/1 (1996), p. 36.

12. Patrick White, *The Tree of Man* (Ringwood: 1955), p. 183.

CHAPTER 2

The nature of Australia

Eric Rolls

Leaving Gondwanaland

Australia has suffered four major disruptions, each of them essential to its present state. The first occurred 75 million years ago when it was stationed near the South Pole as part of Gondwanaland. It snapped away to begin life on its own. There are wondrous geological relics of its life with the rest of the world. At Mount Narryer in Western Australia, one of the earth's shields has come to the surface, a first crust formed 3,800 million years ago. It is a profound experience to touch those rocks; one feels the whole of life. In the Hale River in the Centre, rushing water sculpted a mural of an event that took place 320 million years ago. A cliff thirty metres high and half a kilometre long records it. The ground heaved up over a big area; huge slabs of rock slid one on the other; molten greenschist ran between them, acted as a lubricant and the speed of the slide accelerated. When soft rock slammed against hard rock, it bent, curved, rolled into coils. One can see it all in full colour.

Australia broke away as a geologically quiescent flattish plain covered with rainforest dominated by the ancestors of hoop pine (*Araucaria*) and Antarctic beech (*Nothofagus*). Some living dinosaurs were isolated, but they were all dead when Australia decided to sail north about 30 million years later. The insects aboard were established as they are now. New plants evolved on the journey, including banksias, which reached their modern form 45 million years ago. Birds also evolved, reaching their present form about 20 million years ago. The wetlands of central Australia were lively with cormorants, pelicans, ducks, geese, gulls, rails, cranes and flamingos in great numbers.

Various upheavals altered the face of the land; lava flowed. The surface became complex, the Great Dividing Range was defined down the east. In its present form, which is by no means final, the range is about 6 million years old. It is still rising in places, still subsiding in places, still weathering on top. Compress what happened over millions of years into minutes and the range will writhe like a caterpillar. By the time Australia was nearing its present latitude, it had rotated almost 90°. Fifteen million years ago it brushed against what is now Indonesia and continued its relentless pressure over millions of years. Papua New Guinea acted as a bumper bar and crumpled, pushing up the massive mountain ranges. So our plants and animals had the first influx of foreigners

from the north, among them bats and rodents. Aridity became the normal thing
for a time. The inland seas began to dry up. Eucalypts that had grown on ridges
extended into dying rainforests, casuarinas succeeded auracarias, grasses
replaced ferns, moss and fungi, but there were still huge areas of rainforest and
mixed pasture in districts of higher rainfall.

The arrival of humans

The second disruption to this land that designed itself was the arrival of humans
determined to redesign it to suit themselves. The first arrivals might have
appeared 150,000 years ago. By studying pollens and carbon trapped in coral
reefs, scientists at the James Cook University of North Queensland found a
change in carbon accumulation at that time, suggesting more regular fires. Dr
Gurdup Singh of the Australian National University took core samples from
Lake George, a big lake near Canberra, and obtained pollen and carbon readings
back to 350,000 years. He found a sudden enormous increase in carbon 120,000
years ago.

Where could these people have come from? In April 1996 Dr Neves, Professor of Biological Anthropology at the University of São Paulo in Brazil,
provided substantial evidence. By analysing hundreds of skulls on a computer,
he identified people of Australian Aboriginal appearance who once inhabited
most of east and south-east Asia. When a Mongoloid race came down from
the north about 20,000 years ago, some of the Asian Aboriginal people, who
were expert seamen, island-hopped up through the Philippines, Japan and the
Aleutian Ridge and became the first settlers of the Americas as well as Australia.
Frequent return visits to northern Australia, probably by the Mongoloid people
now named Lapita, began about 5,000 years ago. The dogs they had with them
as pragmatic pets – they also served as food – picked up a species of sucking
louse that evolved on kangaroos and spread it to dogs throughout Asia. Numbers of different people from different places also came to stay, giving great
variance to the Aboriginal people. There was even a pygmy race in north
Queensland; the Tasmanians, who were probably the originals, were different
to all others.

Island seamen are masters at reading the stars; they can tell where they are
by tasting the water – all seas have a different flavour – as well as by reading
patterns of waves and swells not visible to European eyes. The Lapita built
sturdy sailing craft with built-up sides and outriggers as stabilizers. Before then
– and long thereafter – other voyagers had built the sort of vessel that Tim
Severin described in *The China Voyage*, tiers of big curved bamboo logs
strapped together to make a hull driven by big rattan sails, with living-quarters
well clear of the deck. These vessels did not rise and fall with the waves; they
floated steadily while the water poured through them and over them.

The trade in cinnamon gives a good idea of the capacity of early sea-goers.
By 1,000 BC and possibly earlier, Indonesians were travelling across the Indian

Ocean to Madagascar, north to the Gulf of Aden, then down to Ethiopia, where the best cinnamon was grown. They sold it in Rhapta, a busy trading city on the African mainland near the island of Zanzibar, then returned home on the regular yearly change of wind. There and back the journey was 21,000 kilometres (13,000 miles) and took five years.

The different groups of people in Australia developed one common attribute: expert use of fire. The land to which they came was still heavily timbered, not at all the sort of country in which to hunt game. They needed meat as well as the plentiful fruit and greens, but all the game that they could find easily was uncomfortably big: kangaroos were 2 metres tall, a flightless bird weighed 450 kilograms; diprotodons – wombat-like marsupials – weighed 2 tonnes. I do not think there is any doubt, as Tim Flannery points out in his superb book *The Future Eaters*, that the first arrivals hunted them to extinction at the same time as a drying climate made things difficult for these big animals. Spears and boomerangs are not effective weapons against big game. One can readily imagine the same waste when they hunted cattle: cutting off a small mob, they hurled spears, perhaps bringing down one animal while ten or more galloped off to die in the hills with spears through their bellies. By the time smaller animals took over, however, Aborigines had learnt to husband game. Apart from killing only what was needed, they devised systems of taboo, forbidding certain foods to certain people as a method of control.

Fire was a general agricultural tool. By strategic burning to attract animals to new-sprung grass, they always knew where game was to be found. They also used fire to encourage the fruiting of some plants. People in Arnhem Land still extend the fruiting of cycads, a staple food, from three weeks to three months by controlled burning. Like their ancestors, they appoint fire-lighters with flaming torches and beaters carrying leafy branches to put the fire out when it has burnt enough. In the highly flammable heath country of Western Australia, they went to extreme trouble to limit what was burnt. A hundred people spent days breaking down shrubs and throwing them inwards to make a fire-break around a big area shaped like a horseshoe. Then men, women and children stationed themselves with waddies at the wide end, runners lit the curving section and the animals were knocked over as they tried to escape. By intelligence and minute observation, they modified the land to suit themselves. And they made a wondrous job of it. They established a mosaic of growth, varying from months to years after the last fire, until every animal and plant came to terms with fire, either to depend on it or to avoid it. It is a wonderful story that Stephen Pyne told in *Burning Bush*.

It is only recently that we have learnt much of the sophistication of Aboriginal life. Where the climate was unkind, they put up good houses. In wet north Queensland they built domes of latticed cane 3.6 metres in diameter and 2.4 metres high, covered with thatched layers of grass and lined with paper-bark. Tasmanians built to withstand the gales of the south-west, bending down living branches and saplings in a semicircular frame which was thatched with grass,

lined with tea-tree bark and decorated with parrot feathers. In the Centre all
that was needed were shelters from the sun, occasional rain and heavy frosts
that could be built quickly after a seventy-kilometre walk to new water. As
frames, they had to make do with the twisted branches of acacias and stunted
eucalypts, but they fitted them together with such clever use of forks and little
projecting branches that some of them were still standing a hundred years later.
Against this frame they leant a few green branches, then they covered it all
with big balls of spinifex.

They built stone fish traps in rivers and sea and an extraordinary network
of stone mazes to trap eels in western Victoria. When they caught more than
they could eat, they blocked the pens off and fed the fish until they needed
them. They dried meat and fruit; they yarded young pelicans in pens in the
breeding season and the parent birds kept them fed and fat until they were
killed. They mined red ochre as a cosmetic and a paint, and trading tracks to
the main mines ran all over Australia. One group from up near the Gulf of
Carpentaria sent down a hundred carriers each year to a big mine in the Flinders
Ranges of South Australia. That march was 1,600 kilometres, almost 1,000 miles,
one way. They loaded up with 27 kilograms each of ochre, parcelled in grass
or bark, and walked back.

Aborigines lived through astounding changes in Australia. They knew central
Australia as rainforest and freshwater lakes. The Centre began to dry up about
30,000 years ago and, over the next 12,000 years, these dwellers in lush rainforest
became masters of arid conditions far more severe than they are now. The
original population was substantial. The first estimate of 300,000 is ridiculous.
By doing some pretty rough figures, working from an area such as the north
coast of New South Wales, where numbers have been fairly reliably recorded,
then adding that number to a comparable area of equally rich land where
numbers are unknown, and then working like that over the whole of Australia
while allowing for the scant population of the central dry lands, I get the
number up to at least 1.5 million. The greatest error of the early Europeans
was to dismiss these remarkable people as ignorant savages. If they had asked
their advice instead of shooting and poisoning them, Australia would now be
an infinitely greater country.

The advent of the Europeans

The third disruption – this was calamitous, although I must say that I am glad
it happened – was the advent of Europeans to what had become a stable land.
The soil had a mulch of thousands of years. The surface was so loose you could
rake it through the fingers. No wheel had marked it, no leather heel, no cloven
hoof – every mammal, humans included, had walked on padded feet. Digging
sticks had prodded it, but no steel shovel had ever turned a full sod. Our
big animals did not make trails. Hopping kangaroos usually move in scattered company, not in damaging single file like sheep and cattle. The plentiful

wombats each maintained several burrows, so there were no well-used runs radiating from one centre as from a rabbit warren. Every grass-eating mammal had two sets of sharp teeth to make a clean bite. No other land had been treated so gently.

What did the country look like when the white settlers arrived? It looked superb. 'New South Wales is a wildflower garden', wrote Watkin Tench, who came with the First Fleet. 'The country is very romantic, beautifully formed by nature', wrote Elizabeth Marsden, wife of the Reverend Samuel, in 1794. Henry Turnbull, out with the explorer Leichhardt in 1847, was the first to describe the Springsure-Emerald country in central Queensland:

> For upwards of 100 miles we passed through a most splendid, open country consisting of plains and downs – plains stretching out as far as the eye could reach on one side, and beautiful grassy slopes running down from a long and high range of mountains on the other. With only a tree to be seen every 500 or 600 yards, the whole face of the country was covered with the finest grasses and richest herbage, with wildflowers of every tint and colour.

These descriptions of those who saw Australia first are important. I have found many, some still in manuscript, and nearly all show their admiration for the country. That attitude has been lost, clouded over by later literary assessments of men like the poet Adam Lindsay Gordon, who was a bad observer, or Henry Lawson, who was a brilliant short story writer but who saw the land through the eyes of settlers on insufficient areas of land, trying to make a go of it with money borrowed at 20 per cent.

The commonest remark of those practical white men who saw Australia first was: 'You won't have to clear it to cultivate it.' Australia is not a timbered land that has been cleared; that is a common misconception. 'Everywhere we have an open woodland', wrote Charles Darwin when he came there in 1836. 'Nowhere are there any dense forests like those of North America', explained *Chambers's Information for the People* in 1841. The original grasses were mostly deep-rooted perennials that grew in distinct clumps. They had to be deep-rooted because of the nature of the topsoil. Among them grew tap-rooted herbs and many other plants. But always there were bare spaces. Pigeons in great numbers and several species of good-looking native rats had plenty of room to walk about while they fed on fruit and shoots or stretched upwards for the oat-like and millet-like seeds of the grasses. Under the new management, this lovely pasture lasted about six years in most districts. The method of stock management hastened the destruction. Stockmen mustered their cattle in the late afternoon and camped them down on water; shepherds drove their sheep into brush yards and drove them out again each early morning. The ground powdered under the cutting hooves, then hardened when it rained. The plants had never had to push their roots through hard ground; they had never had their leaves bruised by cloven hooves; they had never had whole bunches of leaves torn off between a set of bottom teeth and a top-jaw pad. They died.

Bare ground ringed out from camps and yards. When the stock had to travel too far to feed, yards and camps were shifted; more ground was bared.

Ground never stays bare for long. Inferior Australian grasses, with vicious sharp seeds that corkscrewed into wool and flesh, found the new conditions ideal. They had grown sparsely on rocky hillsides; now they leapt down and took over the flats without competition. Imported weeds with thousands of years' experience of hard soil joined them enthusiastically. It was the kindest climate that they had ever known. With European disruption to Aboriginal burning, trees began to spring up densely where trees had never grown.

What manner of people caused this destruction? They were not greedy and ignorant, as is too often stated; many of them had a background of hundreds of years of good British farming. At home they could estimate pasture, its stocking rate and recovery time, but it was beyond human achievement to assess Australia correctly. It was more a new planet than a new continent. Judging it was even more difficult in low-rainfall country; no one knew what the rainfall was, and areas of 100 millimetres a year looked exactly the same as areas of 200 millimetres. The only difference was in the recovery period after grazing.

Blaxland, Lawson and Wentworth could not have found their way over the great barrier of the Blue Mountains and opened up the interior to settlers in 1813 if the country had carried the present dense growth of tall eucalypts. One can no longer look across a valley to see where a scalable spur begins. Some of the heaviest forest of Victoria's magnificent Gippsland was open country at the time of settlement. When the rather rascally Lieutenant Jeffreys drove four-in-hand from Hobart to Launceston in Tasmania in 1814, before there was any proper road, he saw luxuriant pasture all the way and said there were few trees to impede his progress. Henry Hellyer named the Hampshire and Surrey Hills in north-west Tasmania after the grassy rolling downs that he knew at home; they are now dense forest. Cook's Grassy Hill at Cooktown, where he repaired *Endeavour*, has become scrubby hill. If one goes into the densely timbered country east of Tenterfield in northern New South Wales and traces the long aqueducts cut by Chinese miners in the last century, one realizes that, if the present growth had been there, the work would have been impossible. They could not have seen far enough ahead to peg their lines; roots would have made the digging impractical. On the Palmer River in north Queensland, early gold wardens and geologists stressed the shortage of timber for mine props, boiler fires, even for camp cooking. The country now carries a thousand trees to the hectare. Present Australia grows many more trees than at the time of white settlement. Unfortunately, the growth is disorganized. There are far too many trees where there ought to be few, such as a long wide strip of ruined grassland north of Cobar in New South Wales; there are no trees where there ought to be many, such as the recharge areas of Victoria's Mallee, a huge area of little multi-stemmed eucalypts with bulky lignotubers known as mallee roots. Sometimes they weigh a couple of tonnes. Water that the trees mopped up now pours underground into aqueducts that were never meant – I like to say 'designed' – to carry great amounts of water. That country allowed one tenth of one millimetre of rain a year to soak underground.

After a slow establishment, the rabbits that had been brought in for sport bred in millions. In the hot dry summer of 1890 in the Riverina area of New South Wales, hundreds of thousands would eat a fenced paddock bare, drink all the water in the ground tank, then run in search of food and water. The leaders would crowd against the netting fences, with those coming behind piling on top. When there was a ramp of smothered rabbits, the rest ran over the fence and on towards water. They chewed down kurrajong trees with trunks eighteen centimetres in diameter, then ate leaves, branches and most of the trunk. They ate the bark off big eucalypts and killed the trees. Over a huge area, the pasture became plants unattractive to rabbits. Foxes were brought in for hunting – it was thought to be somewhat undignified to chase wallabies and dingoes – and the first foxes were so precious that they were fitted with collars so that the masters could make a quick snatch at the end and save the quarry for another run. They are now in such numbers in all but the extreme tropical north that they threaten the extinction of numbers of native animals.

Bodies called 'Acclimatization Societies' went into business in the 1860s to distribute animals round the world. Australia sent magpies to England; England sent nightingales to Australia. And they sent starlings, sparrows, blackbirds, hares, deer, trout, skylarks, creatures that thrived and did not thrive. Every ship's cook had a menagerie in his galley. Damage by some of these introductions is only now becoming serious. In many towns and cities, Indian mynas are killing out native birds both by direct attack and by usurping food and breeding grounds. Because of the general derangement of natural conditions, a native bird, the noisy miner (or soldier bird), is having a similar effect. Many species are now in trouble for the first time.

The ruthless European view that almost every natural feature could be improved by engineering led to the organization of river mouths with what are called 'training walls'. Few rivers in Australia escaped such interference. It has interfered with the movement of sand along the coast and in and out of estuaries, resulting in the stripping of beaches and the formation of sandbanks in formerly deep water. In the belief that irrigation would make deserts bloom, river after river was dammed to supply the water. The scarce water was spread too lavishly, and hundreds of thousands of hectares now glisten with the salt brought to the surface by rising water-tables. Swamps that acted as filters for floodwater have been drained in coastal and inland river valleys, resulting in gross pollution of all streams. It is not too late to make corrections; the knowledge is available. Our greatest assets are the children who are learning correct methods at school, especially in the big irrigation districts.

Multiculturalism and the postwar era

The major event of the 1950s was the destruction of more than 90 per cent of
Australia's rabbits by myxomatosis. Plants and animals that had been suppressed
by millions of rabbits for seventy years sprang to life again. They transformed
physical Australia. Almost simultaneously came the fourth major disruption.
Between 1945 and 1965, 2 million migrants came in from all over the world.
They transformed Australia socially from staid inferiority to an exciting modern
country. Australia had long believed that it needed a bigger population. As
early as 1912 the Million Club had been formed in Sydney to encourage
immigration to New South Wales. By 1919 hopes had been raised to the Millions
Club when, with a population of only 5 million, Australia had to shoulder a
war debt of £700 million.

Australia reconsidered its population again at the end of World War II. The
collapse of Singapore made the government realize that a population of 7 million
was too small to be able to do much about protecting itself or to develop the
secondary industries stimulated by war. But immigrants had to be capable of
making an immediate blend with the unimaginative population. In 1944 the
government considered the introduction of 50,000 European children. It was
thought that children would learn English quickly and settle down to Australian
ways. Ten thousand White Russians waiting in Shanghai for permits to enter
Australia could get no replies to applications made many months before. British
only were wanted to maintain familiar Australia.

Luckily, there were not enough available. The greatest number of people of
one foreign tongue to enter the country were 305,000 Italians. The 50,000
prisoners of war, who had been treated well, took home good stories of
Australia. Greeks moved from the islands in the same way as Chinese moved
from villages and counties. The Greeks did not move in clan groups but as
friends, relations, acquaintances. At times, whole islands moved. A few Chinese
from East Timor, anticipating what was to happen there, came to Darwin in
1974. After the Indonesian invasion of 1975, hundreds of frightened women
and children got to Darwin in the clothes that they stood up in and by whatever
ships they could catch. Carrying babies and paper bags, they came ashore to
the shouts of wharfies telling them to go home. By 1985 there were about 3,000
Timorese in the Northern Territory, most of them Chinese and sufficient to
disorder the long-established routine of Chinese life. They came with little
money, so they were willing to do anything and quickly found jobs on roads,
in factories and as cleaners, prawn-sorters, housemaids, kitchen hands. As soon
as they had saved some money, they opened small family businesses, as
the Chinese have always done. This was an economic threat to the established
Chinese, who were unaccustomed to either rocking boats or having their own
boats rocked. After 16 September 1975, when Papua New Guinea became
independent, Chinese with money began to migrate to Australia. About 17 per
cent of them had some Melanesian ancestry; pidgin was often their first

language. Social ranking was based on business success: the part-Melanesians who had done well were regarded as Chinese; those who had failed were natives.

In 1975 there arrived in Darwin an old fifteen-metre fishing boat with five Vietnamese aboard, the first of those who became known as 'boat people'. The story of these journeys is a genuine saga, a tale of heroic exploits that covered twenty years. These first comers had no trouble getting into Darwin. They tied up next to a fishing boat and told the fishermen that they were from Vietnam and ought to report to the authorities. No one believed them, so they went ashore and fitted in. Altogether 121,000 Vietnamese came to Australia, mostly by air from camps in Asia, though 55 boats brought more than 2,000 to Darwin.

Fifty-six boat people from China suffered adventures seldom exceeded in myth. On 16 January 1992, in a wild area named the Kimberley in the far north-west of Western Australia, Cara Holt, daughter of the owners of King Edward River Station, was knocking down termite mounds on their airstrip so that the mail plane could land. This is a weekly job, as termites rebuild their mounds seventy centimetres high in a few days. Regarded as a small station in a kingdom of cattle, the property covers 300,000 acres. The girl looked up to see a Chinese man and woman walking towards her. Their clothes were torn, their feet bound with rags. The woman kowtowed to Cara, her forehead touching the ground. The 16-year-old girl had appeared as a goddess, their saviour. When she took them to her father, they drew a ship and wrote '56' on it. There are no roads in that area, only bush tracks connecting the few stations and Aboriginal communities. Chris Holt went looking for the others in his utility and found thirty. The only English word that they knew was Wyndham, the main town of the area, 200 kilometres east. Another thirteen Chinese walked into the station that afternoon. A grid search by helicopter and aeroplanes located the others in the next few days; ground parties rescued them.

Chen Xin Liang, a primary schoolteacher from Guangzhou, told their story. He had been unable to resist denouncing the government for its increasingly hard Communist line. When agents began looking for him, he went into hiding until relatives told him that some southern Chinese were organising a boat to sail to freedom. The hull was old, the engine older. One day out of China, they hit a storm which holed the boat so badly that they expected to sink. They organized three parties – one to bail, another to patch the hull and the other to work on the engine – and set a course for Vietnam to do repairs. On arrival, they were thrown into gaol. After eighteen days the Vietnamese authorities believed their story that they only wanted to carry out repairs and let them out, giving them oil for the engine. Turned away from Singapore with food, water and oil, they headed for Jakarta and seventeen more days in gaol. Setting off again, they ran into a storm just off Sumba, an island west of Timor. The old boat holed and drifted on to a reef. The islanders got them all ashore, where they were gaoled for four months. Indonesian Chinese interceded with the authorities, bought them another boat and pointed the way to Australia. Having missed Cambridge Gulf and their destination of Wyndham, they tried to thread through the maze of jagged reefs into Montague Sound, holed the boat and ran it ashore on mud-flats. They spent five days vainly trying to repair it, then decided to walk. They had rice, biscuits and dried fish, water in various cans and bottles, compasses and World War II maps.

Although the temperature was up to 45 °C, the weather made the trek possible. Normally, the wet season would have set creeks and rivers running at impassable levels. That year it was particularly late, but lucky storms provided drinking-water. They made fishing-hooks from nails, lines from wire and unbraided nylon string, and caught good fish, one of them 1.5 metres long. They caught goannas, killed a bustard with a fire-hardened wooden spear and clubbed a one-metre crocodile to death. They sharpened a spoon to make a knife to cut it up. One would have thought that such proficient survivors would have been ideal citizens, but none of them found easy acceptance. It took hunger strikes and court inquiries to get some accepted; others were returned to China.

Although immigrants are encouraged and there is a regular intake of 80,000 to nearly 200,000 per year, no government makes plans for them. A yearly intake of 100,000 requires the building of a new Canberra or a new Hobart every two years. The general confusion is highlighted by figures for one period which showed that while immigration increased by 100 per cent, funds for housing were cut by 40 per cent. Nevertheless, Australia changed from official and offensive whiteness to a glorious mix of colours. Modern Australia escaped the controls of ignorant government. If different races had been subjugated into an amorphous English-speaking mass, Australia would have lost substantial blocks of her brain cells. Peter James, headmaster of one of Sydney's govern-ment high schools in the 1990s, encourages students to excel in their own languages. He selects only those teachers who speak several languages. The world is taking an interest in his methods; teachers from many countries came to see the school. James believes that excellence in one language leads to excellence in others – and what better way to begin a study than with one's own language.

In the 1990s more than 4 million Australians had been born overseas, about one in four of a population of 17.7 million; 2.3 million speak a language other than English. In the Sydney suburb of Fairfield alone, 103 languages are spoken; that still tends to be regarded as an aberration instead of a valuable resource. Many of these people have dispersed through the suburbs; many have formed separate communities in all the capital cities, especially in Sydney, where even English, Irish, Scots and New Zealanders have clubbed together. There are at least 3,500 small migrant-owned businesses in Australia, up to one third of which export on a large scale. Migrants have personal networks, not formal business contacts, and they export to their countries of origin. David Kim, a Korean immigrant living in Seven Hills, Sydney, was astonished to find that, in all Australian abattoirs, the feet, legs and some tails of cattle were thrown down the meatmeal chute. Knowing how much his countrymen valued this offal, he formed a successful company to market them. Such immigrants are vital to business and society, but they increase the urbanization of Australia, already perhaps the most urbanized country.

The huge area of the inland, a slab of country equal in size to more than half of the whole of Europe, is unknown to the majority of Australians, new and old. To them, it is more distant and more foreign than Europe. There are now too few to care for it. Athel trees, *Tamarix aphylla*, recommended in the 1950s by the Forestry Commission of New South Wales for planting as shade trees, escaped from somebody's cattle-run on the lonely Finke River in central Australia and choked out 300 kilometres of native plants and their associated animal life in its sandy channel. Feral camels are eradicating parakeelya and pigweed, beautiful succulent plants of the so-called Simpson Desert, no desert at all but originally a dryland wonder. The sand is blowing again for the first time in 30,000 years. Successive waves of rabbits have devastated sandhill shrubs; various exotic weeds are spreading.

It is not too late to save this country. Most of the answers are known. But unless governments take note of what is happening and do something, 75 million years of private creation have been wasted. Always – and this rule is unbreakable – the future must incorporate the past.

CHAPTER 3

The fate of empire in low- and high-energy ecosystems

Timothy F. Flannery

By the beginning of the nineteenth century, the expansion of British imperialism had resulted in the establishment of settler societies on all the habitable continents. The environmental conditions that the settlers encountered were highly varied, encompassing a greater variety of soil and climatic conditions than are present in Europe. This chapter examines the fate of British settlements in Australia and (much more briefly) North America in the light of local environmental conditions.

In the nineteenth and early twentieth centuries, the prevailing view was that the colonial outposts of Australia would coalesce and grow to form a great nation. Comparisons were inevitably drawn with the United States of America, because of the similarities in size, history, ethnic origins and early economies of the two nations. Politicians, journalists and visitors all believed that Australia was a nascent, antipodean United States, just awaiting the fertilizer of population to thrust it into greatness. Despite the prevailing view, there has long been a minority of dissenters, mainly biologists and geographers (Charles Darwin amongst them), who have argued that the environment would dictate otherwise.

These early European settlements of the USA and Australia began as imperial outposts in very different environments. Australian ecosystems are generally characterized by low nutrient availability and highly variable interannual productivity. On the continent-wide scale, it is a low-energy ecosystem.[1] North America is diametrically opposite. There, an enormous extent of extremely fertile soils exists in a region of reliable seasonal productivity. On the continental scale, North America is the archetypal high-energy ecosystem.

Historical ecology and cultural evolution

The fates of the settler societies of North America and Australia are examined here through the lens of historical ecology. The discipline of historical ecology provides valuable insights into the way in which species and cultures are shaped by the evolutionary process. Organisms and cultures, it is postulated, are shaped by similar environmental factors, but they evolve using different mechanisms. Organisms evolve through Darwinian evolution and natural selection acting upon genes, while cultures evolve through Lamarckian evolution acting upon beliefs and practices.

One important aspect of historical ecology is the identification of the principal forces which affect evolutionary direction in different environments. As such, it is concerned with (in order of increasing complexity):

1. the ordering of matter in the non-organic universe;
2. the unique properties of life; and
3. the special properties of human societies, with their economic and political systems.

All of these are sensitive to temporal and spatial variation in energy flows. They are often non-linear in their response, forming a series of steady states punctuated by abrupt transitions (e.g. liquid–gas, heathland–rainforest, hunter-gatherer–agriculturalist). My historico-ecological approach makes several predictions, the validity of which can be tested through analysis of the historical data. They are that:

1. In the long term, the fate of a colonizing society will be determined by the ecology of the land in which it establishes.
2. The degree of difference between the ecology of the settler's homeland and that of the new environment determines the amount of adaptation that the settler society will undergo.
3. The overall volume of biological productivity, its temporal and spatial distribution, and the robustness of the host ecosystem are important factors in determining the orientation and velocity of adaptational change.

A corollary of these broad principles is that one would expect to see convergence between human cultures (and, indeed, the adaptations of other species) which occupy similar environments, regardless of their origins.

In some circumstances, the worldwide market-place has the power to buffer human societies from change. This is because trade allows nutrients and energy to be brought from elsewhere, thus subsidizing societies which would otherwise have to adjust to the lower levels available within their own environments. The importance of this phenomenon in buffering societies from adaptation to local conditions, and the extent to which it can continue in the long term, remains unclear.

In order to understand the impact of ecology upon the colonial settlements, one needs to understand the basic ecological niche and evolution of humanity, as well as the nature of Australian and North American ecosystems. It is also important to understand the adaptations of indigenous peoples.

The nature and evolution of *Homo sapiens*

Humans and their ancestors evolved in east Africa. This region has been, long term, one of the most nutrient-rich and climatically more dependable parts of the planet. The combination of large size, homeothermy, omnivory and sociality which characterizes humans requires enormous energy inputs to support it. It reflects human origins in a nutrient-rich ecosystem.

Recent research suggests that the earliest members of our genus, along with our more distant relatives, made a living as generalized and rare savanna apes who took a little small game, but did a lot of scavenging and plant harvesting. Until about 100,000 years ago, this basic ecological niche had been occupied by our hominid ancestors for several million years. Indeed, except for the importance of scavenging (which was greater to our ancestors), it is essentially the same ecological niche that is still occupied by our nearest living relatives, the two species of chimpanzees, in the rainforests of Africa today.

By around 100,000 years ago, a definite shift in ecology had occurred. Humans were now able to kill moderate-sized prey, such as eland, with relative ease. Larger, more dangerous prey, such as Cape buffalo, elephants, rhinoceros and wild pigs, were still largely beyond the capacity of humans to hunt. By 50,000 years ago, however, even the very largest of land-based prey were regularly falling victim to human hunters. By 40,000 years ago, the people living in northern Africa were making stone points with tangs (a retouched area at the base, to which a handle could be attached). Although spears, perhaps made entirely of wood, had probably been in existence well before this, the tanged spear tips from north Africa were important, for they represented the development of a sophisticated stone-tipped weapon. Such a weapon made it possible to bring down large animals from a considerable distance.[2]

By 20,000 years ago, people all over the world were regularly hunting big game (where it still existed) and had developed sophisticated and diverse stone armouries. Even the largest and most powerful of land mammals, such as mammoths, were a substantial part of the diet of some human groups. This gradual shift in the ecological niche occupied by humans and their ancestors saw them develop from scavengers and hunters of small game to the principal predators of the most formidable of all land animals. The shift meant that, in little less than 100,000 years, humans had developed an ecological niche that overlapped that of virtually all other land-based carnivores. In addition, the human ecological niche had become broader than that of any other mammalian species, for marine resources were also being used, as were an extraordinary variety of plants. Despite the dramatic nature of these changes, it is important to remember that, in terms of human lifetimes, they had happened imperceptibly slowly. Furthermore, they were part of a continuum of change in the shifting ecological niches of the large mammals of Africa and Asia.

Major determinants of Australian ecosystems

The geological, climatic and ecological history of a continent shapes the productivity, reliability and stability of its environments; these in turn shape the ecology of its flora and fauna. These issues are examined in some detail here in relation to Australia, for without a detailed understanding of the environmental determinants of life, the significance of human cultures must remain obscure.

Geology

Over the Cainozoic era (65 million years ago to the present), Australia has experienced extraordinary geological stability and low topography. Indeed, there has been no significant folding of rocks (which leads to mountain-building) within Australia for the past 100 million years.[3] Volcanoes have been periodically active on the eastern margin of the continent throughout the Cainozoic. Moderately extensive Neogene basalt flows have occurred in western Victoria, north-east and south-east Queensland.[4] Erosion of these basalts has resulted in a few 'islands' of relatively fertile soils in an otherwise infertile region. Furthermore, Australia's low topography and temperate location have determined that the extent of glaciation, even during the Pleistocene glacial maxima, has been small. At the height of the last glacial maximum (25,000–15,000 years ago), glaciers covered a mere 6,000 square kilometres in Tasmania and 50 square kilometres in south-eastern Australia.[5] Their contribution to soil production was minimal.

The absence of recent tectonism, widespread volcanic activity or glaciation in Australia has resulted in old, often skeletal soils which are also frequently deeply leached. In addition, Australian soils generally have low organic-matter content, low water storage capacities and high temperatures.[6] They are also notoriously deficient in plant nutrients. On average, phosphorous levels are 300 parts per million (ppm) in Australian soils, compared to 550 ppm and 650 ppm for American and English soils respectively. Levels of nitrogen are likewise low (0.01–0.05 in the drier regions), as are levels of soil sulphur.[7] Levels of trace elements are also very low; deficiences of manganese, copper, zinc, cobalt, boron and molybdenum have all been identified. As a result, approximately 33 per cent of agricultural land in Australia has been treated with trace-element supplements. Altogether, these various attributes of Australian soils mean that they support limited biological productivity. Perhaps the most important factor concerning the evolution of plants is the low levels of phosphorus, for it may have been the determining factor in the evolution of scleromorphy and the zoogeographic patterning of various native plant communities.[8]

Climate

Australia is unique among the continents for the extent to which it is affected by ENSO. ENSO (the El Niño–Southern Oscillation) brings a very large inter-annual variability in rainfall to Australia and this, along with the effects of scleromorph vegetation, make variability in surface water run-off far greater in Australia than on other continents.[9]

The El Niño phase of the ENSO cycle brings drought to Australia. It begins with a change in sea-surface temperature in the eastern Pacific Ocean. Off the coast of Peru, the temperature at the surface of the normally cold sea begins to rise. Eventually, it can rise to 4°C more than normal. Over the course of a year, this warm water can spread into a huge tongue, extending over 120 metres deep and 8,000 kilometres eastwards across the Pacific at the equator. The warm water comes from the western Pacific, in the vicinity of Australia. The warmer waters of the Pacific are normally kept in this region by the prevailing westerly winds. When the winds weaken, the warm water flows back east.[10] Around Australia, the situation is reversed. There, the coastal water is colder than normal and, therefore, evaporation and cloud formation are decreased. Droughts, sometimes of years' duration, occur. Bushfires are prevalent and wind erosion is increased. Effects of ENSO are felt as far away as India, where the monsoon is delayed. Brazil, Central America and southern Africa can also experience drought. It is only in Australia, however, that almost an entire continent is affected. Eventually, the El Niño phase weakens. Westerly Pacific winds re- establish themselves and warm water once again accumulates around Australia. In Australia the drought is often broken with widespread flooding, which promotes further erosion in the denuded landscape.

The length of the ENSO cycle is remarkably variable, ranging from two to eight years. It is this variability as much as anything else that challenges living things. For example, food storage strategies (such as those developed by small overwintering boreal mammals) are useless, for the periods between productive episodes are so unpredictable and often prolonged. From a biological viewpoint, one of the most interesting questions about ENSO is how long the cycle has been in existence. By examining old weather records, climatologists have established that it has been operating since at least the late eighteenth century. Judging from the way in which the Australian flora and fauna have adapted to it, however, researchers suspect that it has been in operation for very much longer than that, perhaps for millions of years.[11]

Tectonic history

Australia's northward drift has matched a deteriorating world climate throughout the Cainozoic, giving unique stability at the continental level for over 40 million years. This long period of *in situ* evolution has resulted in enormous biodiversity. Australia is considered to be one of only eight 'megadiverse' regions on earth.

Impact upon biota

In combination, these phenomena help to explain why Australian trees lose their bark rather than their leaves, why kangaroos hop, why wombats burrow, and why Australia is a great centre of biodiversity. In short, they have produced a unique, fragile and highly interconnected ecology unlike that of any other continent.

A few particular trends that are characteristic of Australian ecosystems and result from these factors are described below.

Biodiversity

Impoverished soils and extreme climatic variability have, paradoxically, resulted in great biological diversity, particularly in scleromorph plant communities and cold-blooded vertebrates. The maintenance of high biodiversity in low-productivity environments is an apparent paradox which is explained by Tilman.[12] He theorizes that such environments can have higher diversity than more fertile regions because 'exterminator species' (highly efficient, high energy-consuming species) are excluded. Where such 'exterminator species' exist, they exclude many other taxa through competition.

Mammals tend to be small, dispersed and with low species diversity

While low-fertility environments can promote great diversity among plants and cold-blooded organisms, species with high energy demands, such as warm-blooded organisms (particularly large carnivorous ones), are greatly disadvantaged. This has resulted in an unbalanced and impoverished assemblage of larger mammals in Australia.[13] One result of this is that the larger Australian mammals weigh approximately a third as much as their ecological counterparts on other continents.[14] The largest flesh-eating mammal ever to have evolved in Australia (the marsupial lion *Thylacoleo carnifex*) weighed less than a human. Very few Australian mammals are highly social; they remain dispersed (or opportunistically clumped) on the landscape.

When the area of Australia is compared to that of other land masses and the number of bird species is calculated on an area basis, we find that Australia's fauna is roughly equivalent, or a little more rich, than that of other continents. Its mammal fauna, however, when viewed on the same basis, is very small. For example, Australia's 250 species compares unfavourably with the 850 recorded from the admittedly larger North America.[15] It may be that Australia's size, erratic climate and poor resource base make it beneficial to be able to fly to reach resources as quickly as possible after they become available.

Homeothermic carnivores are limited in diversity

The number of species in each of the mammalian carnivore guilds is small; each major guild in Australia is occupied by a single small to medium-sized species. This stands in contrast to the seven or more species, including some very large ones, that comprise the more diverse guilds overseas.[16] For example, Europe has twenty-seven species of extant carnivorous mammals, including five canids and three felids.[17] The largest carnivores are the two species of bear, which can weigh between 100 and 300 kilograms. Even during the Pleistocene, Australia had only seven carnivore species weighing more than a kilogram.

Reptiles occupy the large carnivore niche

During the Cainozoic, the largest Australian carnivores were all cold-blooded. At least three large reptilian carnivores – a 3-metre-long land crocodile, a

7-metre-long varanid, and a 6-metre-long constricting snake – were wide-spread.[18]

Carnivory in plants

The large number of carnivorous plant species present in Australia is attributable to poor soils. Half of the known *Drosera* species (sundews) are indigenous to Australia, as is the monotypic family *Cephalotidae* (Western Australian pitcher-plant) and the genus *Polypompholyx* (bladderworts). All of these plants appear to have turned to carnivory in order to obtain nitrates, phosphates and other plant nutrients.

Scleromorphy

Without doubt, the most pervasive and influential of all adaptations in the Australian flora is scleromorphy. The botanist B. A. Barlow called it 'an expression of uniqueness' and described it thus: 'Many ... major [plant] groups are characterized by relatively small, rigid leaves, by short internodes, and small plant size.' Most eucalypts, banksias, bottle-brushes, ti-trees and a vast number of other non-rainforest plant species exhibit scleromorphy to some extent. It is, according to Barlow, 'the most striking aspect of the autochthonous [Gondwanan] element' of Australia's flora.[19]

There is palaeontological evidence that scleromorphy began to develop at least 50 million years ago in Australia. Scleromorphy is a response to the very low levels of phosphates present in Australian soils. Its manifestations, such as small leaves and small distances between leaves, are the consequence of limitations on plant growth which probably result from the small number of new cells that can be sustained at the plant's growing tip.[20]

Energy conservation in mammals

Perhaps the most striking effects of scleromorphy are seen among the few mammal species that feed upon the leaves of scleromorph trees. The best-known of all such species is the koala (*Phascolarctos cinereus*). Its diet consists entirely of eucalypt leaves, but it is extremely selective as to which leaves it eats, preferring those with the fewest tannins and phenolics and the highest levels of nutrients. The koala really lives on the edge, for its food source is so full of dangerous chemicals and so low in nutrients that it has evolved to restrict its energy needs. It needs to eat relatively little; in fact, it is possibly the greatest energy miser of all mammals. Its slow movements and low rate of reproduction are obvious results of this, but less well-known is the extraordinary koala brain. The brain is one of the greatest energy users of all the organs. In humans, the brain weighs a mere 2 per cent of total body mass, yet it uses approximately 17 per cent of the body's energy while not exercising. Because of its high metabolism, it is no surprise that the koala has made some major reductions in brain volume in order to save energy. The strange thing is that the koala brain is much smaller than the cranial vault that houses it (occupying only 67 per cent of the area). Its hemispheres sit like a pair of shrivelled walnut halves on top of the brain stem, in contact neither with each other nor with the bones of the skull. It is the only mammal on earth with such a brain.[21] The koala's major predators are reptiles and other marsupials. Great intelligence may not have been necessary to outwit them. The koala is clearly an extreme, but marsupials in general are not known for their large brains or outstanding intelligence.

In Australia, marsupials survived while condylarth-like placental mammals apparently became extinct.[22] Precisely the reverse occurred on the other continents (except South America), where productivity is not so limited. These patterns of differential survival may relate to relative energy requirements. Wombats, the closest living relatives of the koala, also have an extremely unusual lifestyle which may have been dictated by the limited resources of the Australian environment. The three extant wombat species are the only large herbivorous mammals anywhere on earth to live in burrows.[23] Normally, herbivores need to range over a wide area to meet their energy demands, and spend large amounts of time feeding. This means that they cannot derive much benefit from burrowing. Wombats are able to benefit from living in burrows by keeping their energy requirements extremely low. They require only a third as much energy and nitrogen as a similar-sized kangaroo and spend long hours in their burrows, where they experience near-constant temperature and humidity. Thus they use as little energy as possible. It may be that wombats have been able to evolve such a unique lifestyle by virtue of long evolutionary selection for low energy requirements.

Kangaroos themselves have also been shaped by selection for low energy requirements. Their most distinctive feature – hopping – is an energy-efficient means of getting about. At low speeds, running and hopping use about the same amount of energy, but at higher speeds hopping is more efficient.[24] This is because the energy of each bound is recaptured in the tendons of the legs when the kangaroo lands – rather like in a pogo stick – and is used to power the next leap. Likewise, the force of each leap pushes the gut downwards, creating a vacuum which pulls air into the lungs. Upon landing, the force is reversed, emptying the lungs. This saves the kangaroo from having to use the chest muscles to breathe.

Diversity of small reptiles

In some parts of the arid zone, up to forty-seven lizard species can inhabit a single sand-dune complex. This is a far higher number of species than is found living together anywhere else on earth. Many of the species eat just one food resource – termites. The scleromorphic vegetation of Australia is simply too poor a food resource to be utilized by large herbivores. Termites played a particularly important role in breaking down plant matter and returning nutrients to the soil in Australian ecosystems. The abundance of termites in Australia has thus led to unequalled opportunities for termite-eaters. Lizards, by virtue of their body form and small size, have been able to exploit this resource in various ways, foraging within subterranean tunnels, in tree limbs, in the bases of grass tussocks and in the open.[25]

Low rates of reproduction

Perhaps the most characteristic feature of Australian fauna is a low rate of reproduction. Reproduction is also often opportunistic, rather than seasonal in nature. An account of what is known about mammal reproduction can be found in Strahan.[26] Native rats and mice (*Muridae*) usually have extremely small litters (often one or two) compared with rats and mice elsewhere. Remarkably, there is also some evidence to suggest that the average clutch size of some introduced bird species is declining as they adapt to Australian conditions.

Helpers at the nest

Many Australian birds have a social structure whereby one or more of the young stay with their parents into adulthood, or relatives help to feed another's young. These individuals forgo the chance to raise young themselves in order to help their parents feed their younger brothers and sisters. Kookaburras, noisy miners and fairy wrens all exhibit this behaviour; indeed, it is extremely widespread, almost characteristic of many Australian birds of Gondwanan origin. Elsewhere, it is an extremely rare strategy: about 85 per cent of all species to exhibit it worldwide are Australian.[27] The strategy is clearly beneficial for species living in low-productivity environments.

Self-sacrifice

An even more extreme reproductive adaptation, probably related to nutritional requirements of the young, is seen in the carnivorous marsupials of the genera *Antechinus*, *Phascogale* and *Dasyurus*.[28] Each year, around September, after a frenzied bout of mating, the males die. For a time, the population is composed solely of pregnant females or females with very small young. The lack of males at this time may be critical to the success of reproduction, for the females do not have to compete with males for food when they face the great energy demands of lactation. Likewise, when the young are weaned, they do not need to compete with males for resources. That this adaptation occurs among the carnivores, some of which are very small (30–50 grams), may relate to their higher trophic level. Remarkably, a small lizard, the mallee dragon (*Amphibolurus fordi*), has developed a similar strategy.[29]

Nomadism

Yet another feature of Australian ecosystems to result directly from their inherent constraints is nomadism. Nearly a third of Australia's bird species are truly nomadic, which is an extraordinarily high percentage in world terms.[30] The red kangaroo (*Macropus rufus*) is also partly nomadic, following the rains that produce the short green pick that it prefers. Of course, Australia's first human settlers remained nomadic.

Aborigines

Aboriginal people have been present in Australia for at least 40,000 years. During that time, their culture has been profoundly shaped by Australia's unique ecology. Human populations in Australia have always been small and dispersed, lower in density than those found on other continents. They have occupied a particularly wide ecological niche, and have been densest (approximately forty times the average density) only where rivers have concentrated nutrients.[31] Aborigines have faced a difficult challenge in adapting to life in Australia, for humans evolved in high-nutrient environments and have large energy demands. Aboriginal cultures have become highly specialized. Indeed, in a number of important aspects (e.g. their profoundly complex marriage rules, reciprocal social obligations, unusual exchange practices and high mobility), they depart to a greater extent from the hypothetical ancestral human condition than any other group. Their material culture, in contrast, has remained relatively simple; their tool kits are limited but highly versatile.

Many adaptations of Aborigines seem to be related specifically to high inter-annual variability: in particular, the need to find refuge from irregular, large-magnitude droughts. Such droughts have the potential to threaten the survival of Aborigines over large areas. Firsthand accounts of the effects of such drought are few, but Sturt describes widespread mortality among the Aborigines of the Macquarie River area during the drought of 1828, which may have been of exceptional severity.[32] Bates also records extraordinary overland treks by Aborigines suffering drought.[33] Indeed, because humans reproduce so slowly, it seems possible that the total Aboriginal population may have been kept low (indeed, below carrying capacity) by such exceptional conditioning. Except during severe droughts, Aborigines needed to spend only a few hours per day engaged in subsistence activities.[34] This strongly suggests that they were living at densities well below the maximum carrying capacity of the land. This may help to explain why their tool kits remained relatively simple (except for their ingenuity in developing multi-purpose instruments), for there would have been little evolutionary pressure to improve food-gathering technology under such conditions.

Investing time in agriculture also makes little sense in such an environment. First because food is usually found readily; secondly, because soils are poor (thus yields are low) and the high inter-annual variability in productivity makes yields uncertain and the accumulation of sufficient stocks to last from one productive period to the next highly unlikely. George Grey has recorded proto-agricultural activity amongst the Aborigines of the south-west – one of the very few regions of Australia to receive reliable annual rainfall.[35]

Aborigines typically spend large amounts of time in planning social events. Such events result in the development of extensive networks of social reciprocity, which may have been effective in limiting mortality during droughts by allowing movements of people into regions less affected by the severe conditions. Material culture was typically co-opted to serve this end: edge-ground stone axes, bifacial points, ochre and pearl shells are all examples of goods which were probably more valuable for the opportunity they offered for trade (and thus social contact) than as tools.

European settlers

The first European settlement in Australia was derived largely from urban British origins. It began as a penal settlement with a rigid social hierarchy. After only 200 years of adaptation, the colonial outposts of Australia have already diverged widely from the imperial, European culture which gave rise to them. The development of Australian 'mateship' and the sense of 'a fair go' strongly parallel the Aboriginal sense of reciprocal obligation and also provide refuge (this time in social solidarity) from natural catastrophe. The development of a less hierarchical, more egalitarian social system, along with many other smaller aesthetic and agricultural adjustments (for example, the imposition of realistic stocking rates, the growth in popularity of native gardens), all represent necessary adaptations to allow survival in low-nutrient, highly variable ecosystems. These trends, I hypothesize, will ultimately result in the development of a unique Australian culture which will allow ecological sustainability.

Australia's population has remained small, and there is a growing body of evidence to suggest that it will not expand much further. Indeed, it may ultimately contract. Curiously, Australia is highly urbanized. This may be a response (at least in part) to the extremely variable energy flows caused by high inter-annual variability. Cities tend to derive their income from a large number of sources (many of which are outside the ecosystem in which they are established) and thus tend to buffer their inhabitants from the irregular energy flows generated by local ecosystems.

The North American environment

My comments on North America are sketchy and, necessarily, highly speculative. North America enjoys seasonally dictated productivity in much the same way as Europe, and inter-annual climatic variability is low. Glaciation, volcanism and tectonic activity resulting in erosion have produced youthful topographies. These conditions have persisted throughout the Cainozoic, but Pleistocene glacial activity has accentuated them greatly.

North American soils are, in general, youthful, fertile and deep, many of them having resulted from erosion by the Laurentian ice sheet (which covered 13.5 million square kilometres only 20,000 years ago) as well as earlier glacial events. Latitudinally, North America has been relatively immobile throughout the Cainozoic. A deteriorating world climate over this period has seen the replacement of largely tropical ecosystems with largely temperate ones. In the Pleistocene, glaciation pushed life from the north of the continent, creating a tabula rasa on which a weedy, robust, simple and loosely interconnected eco-system has subsequently established. Biodiversity is typically low, as is endemism, especially when compared with Australia.

Before megafaunal extinction in the latest Pleistocene, North America sup-ported one of the most diverse assemblages of large mammals of any continent.[36] This enormous diversity of large homeotherms, which includes a huge variety of carnivores (including gigantic, carnivorous bears and sabre-toothed cats), had developed in part because North American ecosystems are richer in nu-trients and more productive than Australian ones. North America differs from Australia (and, to some extent, Europe) in that considerable diversity exists within individual species of large homeotherms, even carnivores. The *Ursus horribilis/middendorfi* complex (grizzly and brown bears), for example, shows considerable morphological variability throughout its distribution. Seventy-four 'species' have been described, although only about a dozen of these are currently recognized as valid taxa.[37] These creatures only migrated to North America from Eurasia in the late Pleistocene. While taxonomic splitting may account for some of this 'diversity', it does serve to illustrate that several (up to twelve) morphologically distinct populations of large bears have been able to evolve rapidly in North America. This differentiation in populations of large, homeo-thermic carvivores suggests the existence of an enormous resource base. It indicates that, in some cases, only a small portion of the continent is required to provide the resources to support a viable population of such energy-hungry creatures.

Indigenous humans in North America

Considerable evidence now suggests that humans first arrived in North America from Asia at the very end of the Pleistocene, some 12,000 years ago. Despite the recency of their occupation, enormous cultural diversity developed in pre-Columbian America. By 1492, a multitude of societal types, from migratory hunter-gatherer to city-state, had come into existence. The contrast with Aus-tralian Aborigines, who, in terms of material culture, were relatively uniform from rainforest to desert, is enormous. It can only be explained, I feel, by the differing magnitude of the resource base of the two continents.

The European settler societies of North America

After about 400 years of adaptation, the colonial outposts of North America have also diverged widely from the ancestral type. The enormous economic resource base acquired by the relatively small, relatively homogeneous founding fathers is probably unique in its magnitude. It required a dramatic social reorganization for resources to be utilized effectively. The modern economic state, with its mass production, mass distribution, and high reliance on tech-nology, was the result.

Modern North American culture parallels the Native American cultures in important ways, not least in its growing diversity. Many regional centres are so rich that they form substantial economies, even on a world scale. In a social sense, they are evolving in their own particular direction, adapting to local conditions. The people of the Midwest have probably never felt so distant from those living in New York, while the Californians feel more than ever that they are a law unto themselves. This diversity threatens to tear modern America apart, and communication technologies (themselves a result of the 'American revolution') are involved in a race against time to prevent this.

Other parallels can be drawn. Potlatch bears a strong resemblance to American philanthropy (again, a feature alien to Australian culture), while social inequality within Aztec and American industrial empires stands at the same extreme in its enormity. Further research is clearly needed in order to determine if these comparisons are valid, and herein lies an important test for the hypotheses generated by studies in historical ecology.

Notes

1. Timothy Fritjof Flannery, *The Future Eaters* (Sydney: 1994).
2. G. Burenhalt (ed.), *The Illustrated Encyclopedia of Humankind* (San Francisco: 1993).
3. J. Veevers, *Phanerozoic Earth History of Australia* (Melbourne: 1986).
4. Frederick Sutherland, 'Cainozoic Vulcanism, Eastern Australia: A Predictive Model Based upon Migration over Multiple "Hotspot" Magma Sources', in M. A. J. Williams, P. de Dekker and A. P. Kershaw (eds), *The Cainozoic History of Australia* (Sydney: 1991), pp. 15–43.
5. J. A. Peterson, 'The Equivocal Extent of Glaciation in the Southeastern Uplands of Australia', *Proceedings of the Royal Society of Victoria*, 84 (1971), pp. 207–12.
6. G. Hubble, R. Isbell and K. Northcote, 'Features of Australian Soils', chapter 3, in *Soils: An Australian Perspective* (1993).
7. C. Williams and M. Raupach, 'Plant Nutrients in Australia', chapter 49, in Hubble, Isbell and Northcote (eds), *Soils*; W. J. Hurditch, J. L. Charley and B. N. Richards, 'Sulphur Cycling in Forests of Fraser Island and Coastal New South Wales', in J. R. Freney and A. D. Nicholson (eds), *Sulphur in Australia* (Canberra: 1980).
8. N. C. W. Beadle, 'Soil Phosphate and its Role in Moulding Segments of the Australian Flora and Vegetation', *Ecology*, 47 (1966), pp. 992–1007.
9. B. L. Finlayson and T. A. McMahon, 'Runoff Variability in Australia', in *Institution of Engineers Australia National Conference Publication*, 91/22 (Melbourne: 1991), pp. 504–11.
10. W. J. Burroughs, *Watching the World's Weather* (Cambridge: 1991).
11. N. Nicholls, 'More on Early ENSO: Evidence from Australian Documentary Sources', *Bulletin of the American Meteorological Society*, 69 (1988), pp. 4–6, and 'The El Niño/Southern Oscillation and Australian Vegetation', *Vegetatio*, 91 (1991), pp. 23–36.
12. D. Tilman, *Resource Competition and Community Structure* (Princeton, NJ: 1982).
13. Flannery, *Future Eaters*.
14. P. Murray, 'The Pleistocene Megafauna of Australia', chapter 24, in Patricia Vickers Rich *et al.* (eds), *Vertebrate Palaeontology of Australasia* (Melbourne: 1991), pp. 1074–160.
15. R. L. Strahan, *The Complete Book of Australian Mammals* (Sydney: 1983); E. R. Hall and K. R. Kelson, *The Mammals of North America* (New York: 1959).
16. T. F. Flannery, 'The Mystery of the Meganesian Meateaters', *Australian Natural History*, 23 (1989), pp. 722–9.
17. D. W. McDonald and P. Barrett, *Mammals of Britain and Europe* (London: 1993).

18. Flannery, 'The Mystery of the Meganesian Meateaters'.
19. Barlow, as quoted in A. S. George, 'The Background to the Flora of Australia', in Robertson *et al.* (eds), *Flora of Australia* (Canberra: 1981), pp. 3–25.
20. M. E. White, *After the Greening: The Browning of Australia* (Sydney: 1994), p. 288; Beadle, 'Soil Phosphate'.
21. J. R. Haight and J. E. Nelson, 'A Brain that Doesn't Fit its Skull', in M. Archer (ed.), *Possums and Opossums: Studies in Evolution* (Chipping Norton: 1987), pp. 331–52.
22. H. Godthelp *et al.*, 'Earliest Known Australian Tertiary Mammal Fauna', *Nature*, 365 (1992), pp. 514–16.
23. P. S. Barboza, I. D. Hume and J. V. Nolan, 'Nitrogen Metabolism and Requirements of Nitrogen and Energy in Wombats (*Marsupialia: Vombatidae*)', *Physiological Zoology*, 86 (1993), pp. 807–28.
24. T. J. Dawson, 'Kangaroos', *Scientific American*, 237 (1977), pp. 78–89.
25. E. R. Pianka, *Ecology and Natural History of Desert Lizards* (Princeton, NJ: 1987); S. R. Morton and E. D. James, 'The Diversity and Abundance of Lizards in Arid Australia', *American Naturalist*, 132 (1992), pp. 237–56.
26. Strahan, *Complete Book of Australian Mammals*.
27. H. A. Ford, *Ecology of Birds: An Australian Perspective* (Chipping Norton: 1989).
28. Ford, *Ecology of Birds*.
29. H. G. Cogger, 'Reproductive Cycles, Fat Body Cycles and Socio-sexual Behaviour in the Mallee Dragon *Amphibolurus fordi* (*Lacertilia: Agamidae*)', *Australian Journal of Zoology*, 26 (1978), pp. 653–72.
30. Alan Keast, 'Bird Speciation on the Australian Continent', *Bulletin of the Museum of Comparative Zoology, Harvard*, 123 (1961), pp. 307–495.
31. J. B. Birdsell, 'Some Environmental and Cultural Factors Influencing the Structure of Australian Aboriginal Populations', *American Naturalist*, 87 (1953), pp. 171–207.
32. Charles Sturt, *Two Expeditions into the Interior of Southern Australia*, 2 vols (London: 1833).
33. Daisy Bates, *The Passing of the Aborigines* (London: 1938).
34. This is recorded as being as little as one hour per day by Betty Meehan and Rhys Jones, in S. A. Wild (ed.), *An Aboriginal Ritual of Diplomacy* (Canberra: 1986), pp. 15–32.
35. George Grey, *Journals of Two Expeditions of Discovery in North-West and Western Australia*, vol. 2 (London: 1841).
36. Björn Kurtén and Elaine Anderson, *Pleistocene Mammals of North America* (New York: 1980).
37. William H. Burt and Richard P. Grossenheider, *A Field Guide to the Mammals of America North of Mexico* (Boston: 1976).

PART 2

The Empire of Science

CHAPTER 4

Ecology: a science of empire?

Libby Robin

This book seeks to explore the relationship between two complex phenomena, ecology and empire, in the environmental history of settler societies. Ecology is a way of describing the natural world – it is also a philosophy, an ideology and a science. Ecology/empire interactions build on these multiple meanings of ecology. Ecological imperialism, as described by Alfred Crosby, has already been discussed.[1] A second interaction deals with the ecological limits of empire: the difficulty of establishing European agriculture, pastoralism, and other 'improvements' within pre-existing non-European ecosystems, ecosystems that were themselves dependent on indigenous management techniques. A third interaction is between nature and culture: how did the settlers of the empire come to terms with an 'other' natural world? How did their ecological consciousness, shaped in a 'home' on the other side of the world, adapt to or resist the new environmental conditions?

I will consider ecology here as a science, but in an explicitly political context – as a 'science of empire'. Historians of science, particularly those studying the formation of systematic knowledge at the edges of empire, have elucidated models of intellectual dependency and have traced the changing balance of authority between metropolitan and colonial sources, between 'pure' and 'applied' science.[2] The dependency was not just intellectual but at times explicitly political; imperialism and nationalism continually shaped the ways in which settler-society sciences conceived their local environments. I want to tease out some of the politics of the science of ecology – particularly in Australia, but with brief attention to Britain and the United States – to illustrate the instrumental and strategic association between science and politics, between ecology and empire. An Australian window on environmental history can again alert us to a more general phenomenon: that 'ecology', in its guise as a self-conscious, twentieth-century science, was partly an artefact of empire. 'Empire', too, is a multifaceted term and this chapter will seek to explore the different and overlapping empires that ecological science negotiates in a settler society.

The science of empire and the empire of science

The definition of ecology in Australia was influenced both by the founding political empire of Britain and the emerging scientific empire of the United States of America. The prejudices of both the science of empire and the empire of science have therefore shaped its history. Australian ecology emerged from a background of 'empire science', which began with the sciences of 'exploration': the astronomy of the southern skies; geophysics, including mineral exploration and studies of the earth's magnetism; and natural history, especially taxonomy and systematics. Such sciences offered some sort of key to understanding 'new' lands through the identification of natural resources which might be exploited to justify the great imperial venture. 'Natural resource' benefits were important to the economics of empire. In addition, the British Empire in particular placed a high value on natural history, on 'curiosities'. In Victorian England, the cabinet of curiosities was a mark of class and civilization, a cultural rather than an economic resource of empire.[3]

British plant ecology began as an outgrowth of systematics, its strongest nineteenth-century biological sub-specialty. Systematics itself was a product of the needs of empire for the identification and classification of exotic plants.[4] Scottish botanists William and Robert Smith had begun a regional survey of vegetation in Scotland in 1898. Their survey suggested a way in which the strengths of British systematic work could be merged with new taxonomic vegetation-mapping techniques developed in continental Europe.[5] A. G. (later Sir Arthur) Tansley recognized the value of the model provided by the Smith brothers. In 1904 Tansley set up the British Vegetation Committee to oversee a systematic survey of all Britain. 'Ecology' emerged officially in Britain through the establishment of the British Ecological Society in April 1913, which subsumed the role of the earlier British Vegetation Committee. By this time, Tansley saw the survey process as reaching beyond the basic recording and researching of plant species to investigating patterns of vegetation and the relations between species, including animal–plant relations.[6]

Americans had linked agriculture and 'science for development' very much earlier. Land grant colleges dealt with practical farming, but they also conducted high-level research programmes. At the height of American agricultural optimism in the late nineteenth century, the science of ecology was born in the Midwest, in an agricultural college environment.[7] Some of the prejudices of the 'empire of science' determined that ecology immediately sought to join the ranks of 'pure', rather than 'applied' science. Practitioners perhaps hoped that ecology would assist agricultural science to move beyond 'mere' applied science.

'Ecology' (with its anglicized spelling) was first used at an international botanical congress in Madison, Wisconsin, in 1893 to describe a new sub-specialty in botany. Its distinction was that it used a dynamic perspective to study changes in plant communities. The American school of ecologists incorporated continental European plant sociology and the geography of plant communities to explain the development of changes over time, not merely to describe existing patterns of vegetation. Their dynamic models explicitly analysed the theoretical problem of Darwinian adaptation by plant communities.[8] The new ecologists were keen to demonstrate that their interest in plant communities was philosophical, not management-driven. In this period the term 'ecology' was regarded as technical and obscure and was used only by a small group of professional botanists. One reader of *Science* wrote in 1902 to demand the meaning of the new term.[9] By 1920, when *Ecology*, the American Ecological Society's new journal, first appeared, ecological science was broadening; animal ecology was developing and the president of the society, a forester, advocated ecology as applied interdisciplinary biology.[10]

In Australia, as the nineteenth century wore on, the sciences of exploration gave way to the sciences of 'settling'. The sciences that dealt with the ecological limits of imperial development, especially tropical medicine, agricultural and veterinary science and applied entomology, became more relevant. As in many other colonies and dominions in the early twentieth century, the role of botanic gardens was partly subsumed by departments of agriculture.[11] Imperial scientific effort moved away from centralized metropolitan science and towards applied science based in the colonies and dominions. With the sciences of settling, came the rhetoric of 'science for development'. 'Conservation science', which emerged even later, still incorporated the imperial traditions. These phases of applied biology – particularly 'science for development' and 'conservation science' – will now be analysed in the Australian setting.

The sciences of 'settling'

Australia's largest and most prominent scientific organization, the Commonwealth Scientific and Industrial Research Organization or CSIRO, is government-funded. The links between government and science in Australia are very strong, and a large amount of applied or industrial science is government-funded, both by state and federal governments. The story of the foundation of the original CSIRO, the Council for Scientific and Industrial Research, in 1926, is in part scientific, but it is also very much about the images of the new Australian nation, federated in 1901. CSIR was established because the political leaders of the day were convinced that scientific discovery could render valuable economic service to their developing country.[12]

It is one of the achievements of scientists working at the interface between science and government that the rhetoric of 'science for development' persists in Australia. Perhaps there has been too much success – 'science for development' has tended to become the responsibility of government, almost to the exclusion of industry and private philanthropy. In convincing government to fund science, the Australian polity has convinced itself that science is good government, and scientific experts and advisers are respected accordingly. But Australian industry has not always picked up new scientific opportunities, partly because of limited market size. There are frequent stories of international searches for venture capital to market scientific ideas developed through Australian government funding.[13] The idea of a scientific programme directed towards national development is known to be flawed, but, even so, there is still a basic faith that it offers a useful and politically acceptable direction.

While there is a relatively straightforward association between the applications of science and economic prosperity at the political level, the requirement to be 'applied' is deeply resented within the ranks of scientists. The empire of science that has been built on the institutional structures funded through the rhetoric of science for development has different priorities. Within the international enterprise of science, 'pure' or fundamental science is perceived as more interesting than 'applied', and certain sciences, especially physics and chemistry, are regarded as 'purer' than others. Australian scientists, like their counterparts elsewhere in the western world, value the 'cutting edge' of pure science over what is perceived as the less exciting 'applied' science – 'applied' implying that the science is not new, only the application or context.[14]

'Science for development' rhetoric, which tends to tie science to government, has strong roots in British India in the late nineteenth and early twentieth century. Roy MacLeod describes how the 'scientific soldiers' were recruited to the Indian Civil Service in the hope that science might lead India towards self-sufficiency. The long experience of military rule gave weight to the view that centralized expertise in scientific exploration and research was efficient and cost-effective. Efficiency and cost-effectiveness were, however, notions tied to the distant imperial outpost, where the problems were perhaps singularly amenable to scientific solution: epidemic disease, insect pests, low agricultural yields and poor sanitation. The government of India employed far more scientists in the Indian Civil Service than Britain did in the British Civil Service.[15] In India, the British established 'science services' within government: there were ten by the time the Indian Board of Scientific Advice first met in 1903. But this was a form of science that the British undertook abroad; similar developments were not seen at home until World War I.

Agriculture was at the heart of the colonies and dominions of Britain. British images of 'abroad' were inspired by the yeoman farmer ideal, especially as a way of giving fresh opportunities to its large working class. However, leadership in agricultural science came from outside Britain, especially from the USA. Initiatives at Rothamsted Experimental Farm, the leading nineteenth-century agricultural research station in Britain, had stagnated in the last few decades of the nineteenth century. This was partly because of its ageing scientific leadership, but also – and perhaps more importantly – because of the depressed British agricultural prices in the period – depressed, ironically, because of the success of settler agriculture and pastoralism in North America, Argentina, Australia and New Zealand.[16] Food was cheap and seemingly limitless: there was no incentive to invest in agriculture in Britain. While there was a growing interest in science for industry at this time, agricultural science for development was not a priority.

By contrast, agriculture was a key concern of Australians at the time of federation in 1901, relating to both national identity and national interest.[17] Australia took pride in its agricultural and pastoral economy, 'living off the sheep's back'. The idealistic nationalism of the pastoral frontier portrayed by the *Bulletin* writers of the period also gave Australia a part to play in the wider empire. Imperial Britain expected primary production from its far-flung colonies and dominions, while secondary industry (and its concomitant sciences) was seen to be the appropriate focus of the metropolitan effort. Nineteenth-century agricultural colleges in Australia had focused their efforts on rural education and practical farming. But federation brought increasing rhetoric about science for development and the scientific management of agriculture in particular. Chairs in agriculture were created in major Australian universities to encourage scientific management of the nation's key economic commodities.[18]

The period following World War I was a time for renegotiating and re-envisioning the empire. The war had made the need for industrial science in Britain painfully obvious, as access to German science and manufacturing was now denied. In 1915 Britain established a Department of Scientific and Industrial Research (DSIR). Australia's response was to form an Advisory Council of Science and Industry the following year. But the CSIR, which finally emerged after a further decade of wrangling, was not modelled on the DSIR, despite the apparent similarity of the name; it was, in a sense, almost a mirror image. The DSIR's first role was to support manufacturing industry, and its employees were predominantly physicists and chemists. The Australian CSIR was established to support primary industry, and applied biology was central to its mission.

Michael Worboys has described the interwar period as an era of 'defensive imperialism'.[19] The empire was evolving into the Commonwealth, a more equal partnership, and the dominions demanded concessions in recognition of this. Britain's Colonial Office (responsible for the dominions as well until 1925) saw imperial development as being linked to the biological sciences. It contrasted the needs of industrialized Europe for chemists and physicists with the need for biologists in the agricultural economies of colonies and dominions.

The Empire Marketing Board (EMB) was established in 1926 with a large budget to promote trade within the empire. Its initial emphasis was on publicity and marketing empire products at home and abroad, but, with the financial crash of 1929, its own position became insecure. At the height of its fight for survival, the EMB identified itself increasingly as a centre for intra-imperial planning of research. Its publicity budget was slashed while the research budget grew. Although the dominions did not receive much of the funding, Australia received nearly 40 per cent of what was available – over ten times as much as Canada.[20]

CSIR benefited directly from empire support; the Division of Economic Entomology, for example, received 50 per cent of its funds from the EMB in its early years.[21] But the indirect benefit of the imperial interest in dominion science was even more important, if difficult to quantify. The prospect of an EMB research fund provided the Australian government with a new motivation to revamp and restructure its national science organization. CSIR, structured in such a way as to be receptive to grants from the EMB, was a better-funded and more secure organization than its predecessors.[22]

Early applied biology in Australia was, as in India, driven by the problems of pestilence, of the hazards created by the imbalance between the European crops and animals and the Australian environment. Solutions for such problems did not come initially from a centralized scientific community. William Farrer's successful experiments in breeding varieties of short-season wheat with resistance to rust, the plant pathogen, were very important to the politics of 'science for development' which emerged with the new federal government. Australian farmers traditionally mistrusted 'government advice'. But Farrer, who was both a scientist and a farmer, with credibility on both sides, led a national effort in agricultural science. He was perhaps conscious of the political implications of his position when, in 1902, he named his new prime wheat variety 'Federation'.[23]

Ecology in Australia did not have the clear 'beginnings' of a conference, as in the USA, or of an ecological society, as in Britain. Ecological science was long-established by the time that the Ecological Society of Australia first met in 1961 and began publishing *The Australian Journal of Ecology* in 1976, but its origins are obscure. Ecological investigations were not identified as such until the 1920s, when they emerged from under the umbrella of agricultural science.

In 1922 R. S. Adamson, a British ecologist who had been associated with Tansley's British Vegetation Committee, spent six months in the Department of Botany at the University of Adelaide. The professor, Bentley Osborn, was aware of an urgent need to understand the arid lands in South Australia, which were being developed for agriculture. Adamson brought with him enthusiasm and up-to-date knowledge of British ecology. The Osborn–Adamson collaboration led to the establishment of the Koonamore (later the T. G. B. Osborn) Vegetation Reserve, an ecological reserve of *Atriplex-Maireana* shrub steppes, north of Adelaide. The reserve was situated in the arid zone above 'Goyder's Line' (see Chapter 7, below), where rainfall is so unreliable that cultivation is regarded as uneconomic. It was used to investigate the effects of grazing on such ecosystems. Ecological science began as 'economic botany' on the ecological frontier for agriculture.[24]

Osborn left Adelaide in 1928, moving to Sydney and then to England to succeed A. G. Tansley as Sherardian Professor of Botany at Oxford in 1937. His time at Adelaide, however, was sufficient to create the beginnings of an academic 'school' of Australian ecology. The interest of Osborn's successor, Joseph Garnet Wood, ensured that the school grew and flourished, in collaboration with the Waite Agricultural Research Institute. By coupling ecological research to the pressing economic problems in the agricultural sector, the Adelaide school had access to funding through CSIR and state budgets. Theoretical plant ecology came later, often led by students of the Osborn/Wood school.

Plant ecology led animal ecology in Australia, as it had in Britain and America, but the emphasis on the 'applied', practical applications of the science was more dominant. Ecology in Australia emerged not from systematics or plant physiology, but from 'economic botany'.[25] In a sense, each country's ecology emerged from its strongest cognate field, but all began with plants. Vegetation is funda-mental to delimiting an ecological community (or later, ecosystem), not least because plants stay in place, where animals roam.

The appearance of Charles Elton's *Animal Ecology* in 1927 provided a basis on which zoological ecology could build.[26] Australia's animal ecology began with its agricultural and pastoral pests rather than with its distinctive marsupial fauna. An increasing emphasis on quantification in science demanded that the populations to be manipulated be 'statistically significant', so the emphasis was on working with species in plenty which bred often, rather than rare and endangered ones which bred rarely. There was no theoretical school of animal ecology in Australia parallel to the Chicago school of community ecology under W. C. Allee.[27] Just as Australia's plant ecology was nurtured by economic botany, animal ecology was nurtured by economic entomology.

CSIR's Division of Economic Entomology emerged in the wake of one of Australia's most politically successful 'applied biology' stories, brilliantly told by Eric Rolls in *They All Ran Wild*. It was about the biological control of the pest species, prickly pear. The *Cactoblastus* moth, introduced from Argentina, led to the recovery of between 10 and 25 million hectares of prickly-pear-in-fested brigalow country in New South Wales and Queensland. The moth ate through something like 10,000 tonnes of prickly pear in the four years between 1926 and 1930.[28]

Australia's internationally best-known population ecologist, A. J. Nicholson, worked at CSIR's Division of Economic Entomology. In the early 1930s he developed complex mathematical models of the balance of nature, based on host–parasite population relations, which were well known in the Bureau of Animal Population in Oxford. At home, however, he was better known for his 'applied' work as a sheep blowfly expert.[29]

The secretary of the Empire Marketing Board, Stephen Tallents, had an important and friendly correspondence with David Rivett (1885–1961), the first Chief Executive Officer of CSIR. Tallents found Rivett's vision for an 'empire team' in science both persuasive and useful in internal British and imperial politics.[30] This personal friendship made informal links between CSIR and the EMB possible, the sort that do not show up in tables of expenditure. One of these links was Francis Ratcliffe (1904–70), a young animal ecologist, born in British India, employed by the Empire Marketing Board (see Chapter 5, below).[31] Ratcliffe was sent to Australia in 1929–31 to study the biology of the giant fruit-eating bats ('flying fox'), a major pest species for the eastern coast, especially in Queensland. His trip was one of many taken by the young men of England to the ends of empire. He was described by Tallents as 'much improved ... by his time in Australia'.[32] On returning to Britain, Ratcliffe took up a lectureship in Aberdeen. He 'lightened the darkness of the northern Scottish winters by calling up memories of antipodean warmth and sunshine' by writing a popular memoir of his time in Australia, based on his scientific notebooks.[33] Unlike many of his countrymen, he did not settle at 'home', but looked out for ways in which to fulfil a promise to himself to return to Australia.[34]

Four years later, Ratcliffe seized his chance when CSIR invited him back to report on a different menace, soil erosion affecting vast tracts of outback Australia. In the mid-1930s, the United States was also in the grip of a soil erosion crisis, the famous 'Dust Bowl' of the American Midwest. While Americans called on their distinguished ecologist Frederic Clements, from the Carnegie Institution, to lead a well-funded investigation into the problem, Australians gave the task to a relatively junior British applied biologist. In America, the Dust Bowl served to draw Clements and the whole discipline of plant ecology back into the 'applied' fold after years of theoretical work. In Australia, by contrast, the chosen person had no particular expertise in soil science or plant ecology, but the problem was as extensive in terms of land as in the USA. There were two critical differences: the immediate social cost in sparsely populated Australia was not as high as in the densely populated Midwest; and erosion in Australia was regarded politically as a problem for individual states, not the nation as a whole. Ratcliffe's report to the CSIRO was partially instrumental in founding soil conservation authorities in New South Wales and Victoria in 1938 and 1940 respectively. But the account that Ratcliffe wrote of his outback experiences, *Flying Fox and Drifting Sand*, was to live on in Australian classrooms in the 1950s, raising the consciousness of the next generation of Australians about the ecological limits of agricultural enterprise.[35]

The soil erosion crisis in both the USA and Australia changed the emphasis of applied science. There had been enormous confidence in 'science for development' in both places, but soil erosion could not be handled on a 'pest control' model. It was not a single problem and it required an interdisciplinary approach. There was increasing recognition that one could not simply have a 'Grow More Wheat Year', as had been advocated in Australia after the 1929 financial crash. Science became the voice of reason and restraint, of management for a long-term yield. The conservation science that emerged in the 1930s and 1940s was still about developing the nation, like the earlier 'science for development', but its 'progressive' agenda increasingly emphasized development in the long term, not instant results.

Conservation science, which included forestry and soil conservation, became the next important umbrella for ecological work in Australia. Vegetation mapping in key watershed areas such as the Australian Alps was important in introducing the next generation of botanists and agricultural scientists to the science of ecology. Such work was often partly funded by state soil conservation authorities.[36]

Francis Ratcliffe, after preparing his report on soil erosion in the 1930s, became a permanent member of CSIR's Division of Entomology, where he undertook work on termites and the pests of stored wheat. He also worked on the malaria mosquito during the war. In 1949 he became the first Officer-in-Charge of CSIRO's Wildlife Survey Section, established to undertake a national biological survey. But the applied and agricultural imperative of CSIRO forced the Wildlife Survey Section to concentrate on that most persistent of imported species, the rabbit. Ratcliffe worked closely with the virologist Frank Fenner (1914–) on myxomatosis, the rabbit-specific disease which was dramatically to reduce rabbit populations for many years.

The biological survey work languished. But this was only partly because of the distracting problem of rabbits. The other part of the problem was that the empire of international ecology had moved away from descriptive surveys. Australian ecologists, in line with overseas trends, increasingly focused on quantification, seeing that as the vanguard of 'interesting science'. They felt compelled to distance themselves from anything that smacked of 'bucket science'. The pejorative 'bucket' was regarded as the sort of gross quantificational device that was suited to a science that had not yet reached respectable levels of precision measurement. The comprehensive survey of plant and animal community associations, like the one begun in Britain by A. G. Tansley's British Vegetation Committee in the first decade of the twentieth century, was still missing for Australia. National surveys of indigenous flora and fauna had been advocated by the Australasian Association for the Advancement of Science since the 1880s.[37] But no one had taken up the challenge while it was 'frontier' science and, by the 1950s, it no longer had that status.

The cumulative work of Australian ecologists, plant and animal alike, focused predominantly on the development of non-indigenous food species. Acclimatization and the development of improved varieties of imported species, along with pest, pathogen and weed eradication, were the sciences that were perceived as serving 'national needs'. Indigenous flora and fauna had rarely been the subject of study, and the lack of groundwork in this sort of biology made respectably quantified ecological work even more difficult.

Eight years after Ratcliffe had begun work in the Wildlife Survey Section, he remarked that his department had not been able to collect or collate even basic 'data needed for appraisal of the present situation', or 'information on the status and distribution of the more interesting species'. The section had not made any assessment of the biological adequacy of national parks and reserves, 'in competition ... with [its] normal research activities'. A survey of 'the status of marsupials in the State of New South Wales' was the principal achievement of many years of the section's research effort.[38] Even Australia's strength in entomology was stretched. As Tim Flannery has observed: 'there are more species of ants inhabiting the hill called Black Mountain that overlooks Canberra, than in all of Britain'.[39] The ecological specialty, such as it was, was not in a position to tackle all the indigenous species and ecosystems of the vast continent of Australia.

Despite the fact that ecology was doubly disadvantaged by being both biology and 'applied', it occasionally became the focus of the élite science empire. The Australian Academy of Science (founded in 1954) sought, as one of its first priorities, ways in which to undertake science across disciplines and in the national interest. The Academy wanted to reinforce the rhetoric of the national usefulness of 'science' (perhaps as opposed to particular sciences) and to be qualified to comment on scientific matters at international forums. Ecology in its broader, interdisciplinary sense fitted the Academy's local political needs well, especially its high-profile concern with establishing national parks and conservation reserves.[40] Ratcliffe was quick to realize that his pet project, to develop a policy for Australian wildlife conservation that would be 'planned, positive and national', might fit under the Academy's umbrella. He recognized that this required 'more than a dash of idealism', and in 1957 appealed to the Academy to support and sponsor this goal.[41]

By the 1960s, the weight of international imperatives had enveloped the Australian Academy of Science. Australia's responses to organizations like the International Union for the Conservation of Nature and the International Biological Program (IBP) were filtered through the Academy. The United States co-ordinated a 'Manhattan Project' style of research for IBP ecology (Man and Biosphere). 'Data' from all over the world was collated centrally, on the basis of IBP guidelines. It was the last great imperial exercise in ecology, with information from the periphery being sent to a metropolitan centre to be converted into 'science'.

Australian scientists expressed doubts about IBP. John Turner, one of the Academy's two representatives at the Paris meeting to set up the IBP in 1964, wrote in his report that 'the subject of Biology at present is not yet ripe for an International Programme'.[42] The focus on ecology, however, demanded that Australia participate because its ecosystems were so distinctive. The national pride of the young Academy was also at stake. Turner sought to shape the Australian contribution in a way which would meet the international needs as far as possible, but which would also serve national conservation objectives relevant to the Academy's profile at home, especially the designation of national parks and conservation reserves.

Australian IBP ecologists, led by R. L. Specht (1926–), who was trained in the original Adelaide heathland ecology school, came to the same conclusion as the Americans about the limitations of the IBP enterprise: 'Since the information compiled on individual biomes was not yielding data applicable to other systems, it was not leading to a useful theoretical synthesis.'[43] That elusive theoretical synthesis has dogged the science of ecology throughout its history, adding to the pressure of political imperatives. The lack of 'universal panacea' in the discipline meant that each ecological response, whether to local conservation concerns or to IBP, required individual research.[44] At the height of the environmental revolution in 1969, Turner wrote: 'everything is progressing so quickly that the burden on the few ecologists … is becoming almost intolerable'.[45]

At one stage Ratcliffe had hopes that the emerging political enthusiasm for the environment would provide a new source of funds for science, especially the science that was necessary to mount solid political arguments about ecological issues. Ratcliffe was a key founder of the Australian Conservation

Foundation (ACF), established in 1965, and he worked for it until just a couple of months before his death in 1970. Ratcliffe's vision for the ACF was based on the assumption that conservation was science, and that ecological science was a 'cause' worthy of non-government funding. But development that threatened the environment would not wait for the measured scientific evaluation that Ratcliffe had in mind. The ACF moved quickly to invest more in publicity and politics and left its early scientific initiatives behind.[46] Ecological science was forced back to the domain of government.

Ecology in Australia is still, in a sense, a science of empire. Its funding continues to be tied to 'national priorities'. The empire of economics has survived beyond the British one, and the empire of science still dictates structures for 'good science', especially the search for 'universal solutions', which are deeply problematic for ecology. Australia's most expensive scientific efforts are still directed at acclimatizing imports. The emphasis is on 'improving' flora and fauna known to be economically useful in the short term, rather than on understanding indigenous species irrespective of their long-term economic and ecological potential.

Australian ecology has focused on the species of invasion, the shipboard imports of the Crosby model. Its first tentative steps were taken at the ecological limits of agriculture, beyond Goyder's Line. The consciousness that has shaped ecological studies has been very late to consider distinctively Australian species. Until relatively recently, even *conservation* science has, paradoxically, favoured imported species. The western thinking that is ingrained in science is deeply imperial. It has taken two centuries for biologists to work co-operatively with Aboriginal Australians, to incorporate systematic consideration of indigenous ecological understandings.[47] This line of research is most promising for the release of Australian ecology from its service to empire.

Notes

1. Alfred Crosby, *Ecological Imperialism* (Cambridge: 1986); Tom Griffiths, Introduction, above.
2. The classic intellectual dependency model is George Basalla, 'The Spread of Western Science', *Science*, 156 (1967), pp. 611–22.
3. David Elliston Allen, *The Naturalist in Britain* (London: 1976).
4. I am using 'biology' here as distinct from natural history and medicine: A. G. Morton, *History of Botanical Science* (London: 1981), pp. 364, 405 (footnote 6). See also Michael Worboys, 'Science and British and Colonial Imperialism 1895–1940', PhD thesis (University of Sussex: 1979), p. 305, where it is argued that the dearth of biology graduates drove the Colonial Office to promote biology education in Britain for the colonies.
5. They used techniques learned (by Robert) under Flahault in the Zürich–Montpellier school, which regarded vegetation mapping as a form of taxonomy: Peter J. Bowler, *The Fontana History of the Environmental Sciences* (London: 1992), p. 526.
6. A. G. Tansley, 'The Aims of the New Journal', *Journal of Ecology*, 1 (1913), pp. 1–3, 'Formation of a Committee for the Survey and Study of British Vegetation', *New Phytologist*, 4 (1904), pp. 23–6, 'The Early History of Modern Plant Ecology in Britain', *Journal of Ecology*, 35 (1947), pp. 130–3.

7. Donald Worster, *Nature's Economy* (Cambridge: 1977); Ronald Tobey, *Saving the Prairies* (Berkeley: 1981).

8. Eugene Cittadino, 'Ecology and the Professionalization of Botany in America, 1890–1905', *Studies in the History of Biology*, 4 (1980), pp. 171–98; Ronald Tobey, 'Theoretical Science and Technology in American Ecology', *Technology and Culture*, 17 (1976), pp. 718–28; Richard A. Overfield, 'Charles E. Bessey: The Impact of the "New" Botany on American Agriculture 1880–1910', *Technology and Culture*, 16 (1975), pp. 162–81.

9. Worster, *Nature's Economy*, p. 203.

10. Barrington Moore, 'The Scope of Ecology', *Ecology*, 1/1 (1920), pp. 4–5.

11. Worboys, 'Science and British and Colonial Imperialism', p. 35. See also 'The Imperial Institute: The State and the Development of the Natural Resources of the Colonial Empire, 1887–1923', in John M. MacKenzie (ed.), *Imperialism and the Natural World* (Manchester: 1990), pp. 164–86.

12. George Currie and John Graham, *The Origins of CSIRO: Science and the Commonwealth Government 1901–1926* (Melbourne: 1966), p. v; CSIR became CSIRO in 1949.

13. On markets, see Jan Todd, *Colonial Technology: Science and the Transfer of Innovation to Australia* (Cambridge: 1995), pp. 240–8. In the 1970s, Australia had 0.35 per cent of the world's population, accounted for 1.6 per cent of the world's scientific literature, but only 0.31 per cent of the world's emerging patents. See James Davenport, 'The Impulse of Science in Public Affairs, 1945–1986', in Roy M. MacLeod (ed.), *The Commonwealth of Science* (Melbourne: 1988), pp. 73–96.

14. Worboys, 'Science and British and Colonial Imperialism', p. 227; Roy M. MacLeod and E. Kay Andrews, 'The Origins of the DSIR', *Public Administration*, 48/1 (1970), pp. 23–48, esp. p. 31.

15. Roy M. MacLeod, 'Scientific Advice for British India', *Modern Asian Studies*, 9/3 (1975), pp. 343–84. Compare the British Imperial Forestry Service in India: Stephen J. Pyne, Chapter 1, above.

16. Sir E. John Russell, *A History of Agricultural Science in Great Britain, 1620–1954* (London: 1966), esp. pp. 159–79.

17. Currie and Graham, *The Origins of CSIRO*.

18. The Universities of Sydney, Melbourne and Western Australia appointed chairs in agriculture in 1910, 1911 and 1913 respectively (Robert Dickie Watt, Thomas Cherry and John Waugh Paterson).

19. Worboys, 'Science and British and Colonial Imperialism', pp. 192, 218.

20. Figures based on A. C. D. Rivett, 'The Empire Marketing Board', in *The Australian Rhodes Review* (Melbourne: 1934), pp. 12–14.

21. Rivett, 'Empire Marketing Board', p. 13.

22. G. Currie and J. Graham, 'Growth of Scientific Research in Australia', *Records of the Australian Academy of Science*, 1/3 (1968), pp. 25–35. See also C. B. Schedvin, *Shaping Science and Industry: A History of Australia's CSIR 1926–1949* (Sydney: 1987), p. 17; Currie and Graham, *The Origins of CSIRO*, pp. 106–34.

23. An accessible summary of Farrer's wheat-breeding programme is in C. W. Wrigley, 'William James Farrer (1845–1906)', in Bede Nairn and Geoffrey Serle (eds), *Australian Dictionary of Biography*, vol. 8 (Carlton: 1981), pp. 471–3.

24. Tansley, 'The Early History of Modern Plant Ecology in Britain', p. 135. R. L. Specht, 'Australia', in Edward J. Kormondy and J. Frank McCormick (eds), *Handbook of Contemporary Developments in World Ecology* (Westport: 1981), dubbed Osborn and his successor, J. G. Wood, as 'father and son' of ecology in Australia (p. 408). Goyder's Line was established by the nineteenth-century Surveyor-General, George Goyder, as the limit to arable land in South Australia: J. M. Powell, *An Historical Geography of Modern Australia* (Cambridge: 1988); Michael Williams, *The Making of the South Australian Landscape* (London: 1974), pp. 306–12.

25. Physiological studies in ecology emerged later, under R. N. Robertson (1913–) (Adelaide). J. S. Turner (1908–91) at Melbourne was also a plant physiologist and ecologist.

26. Specht, 'Australia', p. 388, identified Elton's work as critical to Australian ecology.

27. See Gregg Mitman, *The State of Nature: Ecology, Community and American Social Thought: 1900–1950* (Chicago: 1992).

28. Figures adapted from Eric C. Rolls, *They All Ran Wild* (Sydney: 1984), pp. 440–2. Brigalow is the open forest of drier south-eastern Queensland, dominated by *Acacia harpophylla*. On the Commonwealth Prickly Pear Board, see Schedvin, *Shaping Science and Industry*, pp. 90–6.

29. John L. Hopper, 'Opportunities and Handicaps of Antipodean Scientists', *Historical Records of Australian Science*, 7/2 (1988), pp. 179–88, esp. p. 186.

30. A letter from Rivett to Tallents, 4 February 1932, refers to the 'empire team' and to his confidence that the EMB provides the most effective way of managing this (File 10, Tallents Papers, Institute of Commonwealth Studies Archive, London).

31. Biographical details of Ratcliffe's life from 'Leading CSIRO Scientist Dies', Canberra Times, 3 December 1970; 'Top Scientist, Author Dies', *Canberra News*, 2 December 1970; 'Wildlife Research Pioneer Dies at 66', *Australian*, 3 December 1970 (Cuttings File, MS 2493, Ratcliffe, National Library of Australia, Canberra).

32. Tallents to Rivett, 23 September 1931 (File 10, Tallents Papers, ICS).

33. Francis Ratcliffe, *Flying Fox and Drifting Sand: The Adventures of a Biologist in Australia* (1938; Sydney: 1947), p. viii.

34. Ratcliffe, *Flying Fox*, p. 183. The urge to seek scientific opportunities abroad was no doubt at least partly because there were still few opportunities for biologists in Britain: see Worboys, 'Science and British and Colonial Imperialism', p. 305.

35. For more on Ratcliffe and soil conservation, see Thomas R. Dunlap, Chapter 5, below.

36. Libby Robin, 'Radical Ecology and Conservation Science: An Australian Perspective', *Environment and History*, forthcoming.

37. Linden Gillbank, 'The Life Sciences: Collections to Conservation', in MacLeod (ed.), *The Commonwealth of Science*, pp. 99–129.

38. F. N. Ratcliffe, letter to Australian Academy of Science, 1 May 1957, pp. 4–5 (File 1002, 'National Parks', Academy of Science Archives, Canberra). Even the New South Wales survey had required joint funding from the state government.

39. Timothy Fritjof Flannery, *The Future Eaters* (Sydney: 1995), p. 75.

40. Libby Robin, 'Nature Conservation as a National Concern: The Role of the Australian Academy of Science', *Historical Records of Australian Science*, 10/1 (1994), pp. 1–24.

41. Ratcliffe, letter to Australian Academy of Science, 1 May 1957, p. 1.

42. J. S. Turner, draft report on the International Biological Program of the ICSU meeting in Paris, 23–25 July 1964, p. 4 (Box 26, Turner Collection, University of Melbourne Archives, Melbourne). The academy's other representative was the geneticist Sir Otto Frankel (1900–) from CSIRO (Division of Plant Industry).

43. As summarized by Dorothy Nelkin, 'Scientists and Professional Responsibility: The Experience of American Ecologists', *Social Studies of Science*, 7 (1977), p. 85.

44. See Thomas R. Dunlap, Chapter 5, below.

45. Professor John Turner to W. J. (Bill) Kilpatrick of Hawthorn, 17 June 1971 (Box 20, Turner Collection, University of Melbourne).

46. Robin, 'Nature Conservation as a National Concern', pp. 11–16.

47. Peter Latz, *Bushfires and Bushtucker: Aboriginal Plant Use in Central Australia* (Alice Springs: 1995); J. R. W. Reid, J. A. Kerle and S. R. Morton (eds), *Uluru Fauna* (Canberra: 1993).

CHAPTER 5

Ecology and environmentalism in the Anglo settler colonies

Thomas R. Dunlap

Australia, Canada, New Zealand, and the United States are the Anglo part of what Alfred Crosby called the 'neo-Europes' and Geoffrey Bolton called the 'colonies of settlement'.[1] They are the countries, and the only ones, in which English-speaking settlers dispossessed and almost exterminated the earlier inhabitants, establishing a new society and government modelled on the old. In the 'colonies of empire', India being the pre-eminent example, a small Anglo population ruled an overwhelming mass of native peoples; South Africa, with two relatively large minority European populations, is an intermediate case. The 'colonies of settlement' retained their cultural ties to Britain – even the United States, which broke its political ones with considerable force – and built and maintained networks among themselves. Much of the history of the colonies of settlement cannot be fully understood from the perspective of nation or empire. It must be seen in the context of this 'Anglo world'.

The settlers' understanding of nature is one such element. From the first, they have understood their land through their European culture, but in each country and era they have had to apply its concepts to their own situation. Atoms and molecules are everywhere the same. Plants, animals, and the land – the level at which people experience and construct the world – are everywhere distinct. Since the late eighteenth century, the organized nature knowledge of the culture, what the settlers call 'science', has been a primary influence. Here I shall consider one key part of that: the development of ecological ideas in the United States and Great Britain in the interwar period, and their transmission to the scientific communities of Australia and New Zealand in the postwar years (Canada's uniquely close connections to the United States make it a special case which is only considered briefly here). This, the background of environmentalism, shows some of the connections within that world, the inter-play between local learning and the larger cultural context, and the impact of science on policy.

As a 'pure science' dealing with nature as people saw it and cared for it, ecology emerged in the interwar years, most notably with Charles Elton's *Animal Ecology*, which put forward concepts that guided a generation of research.[2] Equally important, though, was research in game management, an applied science that grew up in the United States during these years in response to the crisis of game production. Both areas emphasized a new view of nature and a new way to study it. It was a set of intricately connected systems that could only be understood through quantitative studies of complex interactions among species and with the land. After World War II, this knowledge began, albeit slowly, to shape policy and ideas in Australia and New Zealand. This is in part a conventional tale of ideas transmitted from the metropolis to the periphery, but only in part. It cannot be seen solely as a chapter in 'colonial science'.[3] Rather, it is a study of connections within the Anglo world, of institutions and arrangements not designed to build national scientific communities, but to connect a small national group to a wider circle of knowledge and technique.

The place to begin is the first important game management study, Herbert Stoddard's work on bobwhite quail in Georgia (1924–29), for it shows how the perspective of an interconnected world and a reliance on quantitative methods were developing outside as well as within 'pure' science. In 1923 a group of wealthy sportsmen approached E. W. Nelson, head of the US Department of Agriculture's Bureau of Biological Survey. They were concerned about the declining numbers of birds on their quail plantations and they asked the agency to conduct research, which they would support. They wanted practical advice, and there was nothing in the arrangement to suggest that they might get anything more. The bureau was not notably scientific; Nelson was an old-guard naturalist-collector; and the project's supervisor, W. L. McAtee, was a specialist in analysing the contents of birds' stomachs, a technique pioneered in the 1870s. Stoddard, who did the fieldwork, had dropped out of school at 15 to become a farmhand and had moved into science by collecting and mounting specimens for the Milwaukee Museum. The project, though, was anything but old-fashioned. It was a quantitative study of the environmental factors affecting quail over their life history, the first of its kind in the world. It followed fairly closely the recommendations that Elton was then making for field studies that would 'revolve around censuses, the structure of the population by age and sex, the birth-rates and death rates, movements, as well as the influence on these of outside changes and interrelations'.[4] Elton had written *Animal Ecology* in 1927, but Stoddard did not read it until 1931, when Aldo Leopold loaned him a copy. He sent it back six months later with the non-committal comment: 'I must make a point of getting his books; his comes the nearest to being the sort of ecology I can appreciate.'[5]

Stoddard's study was marked by a new view of nature and the way in which humans might understand it. The common view, popular and scientific, framed the world in terms of interactions on the land, which remained largely a neutral backdrop. Observation and common sense revealed the processes by which land and life were connected. Stoddard, however, believed that the land was an active and effective agent and that the unit of study, therefore, was not the species but the population. He came to conclusions about quail in south Georgia, not about quail in general. He also believed that interactions were complex and could only be understood from rigorous, quantitative observations. Nature was not observed, it was constructed.[6] The common belief, for example, was that hunters and hawks were what kept the quail population down: cut seasons and bag limits, put bounties on hawks, and there would be more quail.[7] By counting nests and eggs and calculating mortality from egg to adult, however, Stoddard found that cotton rats and ants, which destroyed the eggs and young, were more significant than the hawks that killed the adult quail. He also found that the key element was farming practices. Late nineteenth-century farms and country-side, with their dirt and gravel roads, snake fences, small fields, and mixed farming and woodlots, provided ideal conditions for quail. Modern farming was destroying the fence-rows that sheltered them and the weeds that they ate, while blacktop was covering the grit that they needed for their crops and the dust in which they bathed to keep down parasites.[8]

This perspective was not unique to the United States or to game management. It was becoming the common property of field biology, as we can see in the work that Francis Ratcliffe did for Australia's Council for Scientific and Industrial Research (later CSIRO), from 1929 to the early 1940s (see Chapter 4, above). Ratcliffe had more education than Stoddard, but it was not specifically ecological. He had studied at Oxford under Julian Huxley just after World War I, 'one of a band of able and attractive students' (Charles Elton was another) who came to the university 'at a time when the foundations were being laid for that renaissance of general biology' which was so prominent a feature of the 1930s (as Huxley put it).[9] Hired in 1929 to investigate the damage of flying foxes to the orchards of Australia's east coast, he used methods that might have been employed by natural historians a generation before – collecting specimens, tracing migrations, and examining the contents of their stomachs – but his analysis was quite modern. The fruit growers saw the problem as the existence of 'a pestilential animal that required extermination, and the solution a practical method of wholesale destruction'. Ratcliffe had a different view: that his task was to get on 'accurate picture of their population as a whole and what might be called their economy'.[10] He also had a different conclusion. Eradication, he said, was not necessary, desirable, or practical. Only a small minority of the animals did any harm, and they did it only sporadically. All that could or should be done was to shoot particularly troublesome flocks. The problem, in short, was not the species, but the behaviour of a part of the population.

This view is even more apparent in his study of the central Australian drought, where he applied this perspective not to wildlife populations but to human settlers. Faced with a disaster much like that of the Dust Bowl in the United States, CSIR hired Ratcliffe in 1935 to look into the situation. As in Queensland, he worked alone, using conventional methods. He went round the country on a motor bike or hitched rides from mail carriers and graziers, looked at the land, and asked questions. The scale of this effort was in sharp contrast to the resources that the New Deal was then deploying in the United States to study the Dust Bowl, and suggests how differences in resources may shape research. After he had submitted his report, Ratcliffe said, he spent a year mulling over the problems of pasturing sheep in the inland. He found a solution, but 'it was not the answer I wanted or hoped to find'.[11] The land was not suited to the settlers' economy or way of life.

> The essential features of white pastoral settlement – a stable home, a circum-scribed area of land, and a flock or herd maintained on this land year-in and year-out – are a heritage of life in the reliable kindly climate of Europe. In the drought-risky semi-desert Australian inland they tend to make settlement self-destructive.[12]

If the graziers tried to meet the inevitable but unpredictable droughts by selling their flocks, they would lose their investment and their breeding programmes. If they did not, they would ruin the range. The only thing that would preserve the country would be 'consciously to plan a decrease in the density of pastoral population of the inland ... '.[13] The analysis did not appeal to ecological theory and it concentrated on economic and social elements, but it was ecological in the broad sense. It saw the situation in terms of relationships, processes, and limits imposed by the environment, and the solution was to fit the social and economic order to the natural one.

Ratcliffe's work on rabbits during World War II shows these characteristics even more clearly. People had commonly described rabbits as a general phen-omenon – the 'rabbit menace', the 'grey blanket', an invasion, or an epidemic. But Ratcliffe pointed out that the interaction of land conditions and the econ-omics of settler agriculture created three quite different situations. In the comparatively well-watered areas near the coast, where farming was intensive, rabbits were not a major problem. Normal farming operations, some purposeful ploughing-up of warrens, and a bit of trapping, poisoning and shooting kept their numbers down, and profit per acre was high enough for the farmers to be able to afford the extra cost. In the very arid pastures surrounding the interior desert there was no need for control; frequent, harsh droughts 'knocked back' the population. It was in the intermediate areas – unfortunately, the heart of pastoral Australia and much of the wheat belt – that rabbits were a scourge. Here rainfall and plant growth were sufficient to allow the animals to flourish, but the return per acre was too low to bear the added cost of control.[14]

In the United States, two brothers, Olaus and Adolph Murie, were – unintent-ionally – producing an even clearer demonstration of the way in which ecological conclusions did not depend on formal training or research methods. Between 1927 and 1932 the elder brother, Olaus, working in the Bureau of Biological Survey under W. L. McAtee's direction, studied the impact of coyote predation on the elk of Jackson Hole, Wyoming, by examining the food habits of the animal. It was a conventional study of stomach contents, carried out by a man who had been doing similar studies since graduating from college in 1912. As he was finishing, Adolph was starting work on the same subject a few miles north, working for the National Park Service in Yellowstone National

Park. His methods were different from his brother's and the scope of his investigations wider. Like Olaus, he had looked at stomach contents, but he also examined the interactions between coyotes and many other species and reconstructed a picture of animal abundance in the region since the late nineteenth century. His research was as comprehensive as its title, *The Ecology of the Coyote in the Yellowstone*. The reason for these differences was that Adolph was ten years younger. When he had graduated from college, it was possible to do advanced work in animal ecology and he had earned a PhD in the subject at the University of Michigan. However, the brothers approached the problem in the same fashion – as the connections between populations in a defined area – and they came to the same conclusion. Individual animals were vulnerable to predation, populations were not.[15]

By the time that World War II interrupted ecological research and education, scientists and game managers in the United States and Britain had laid the intellectual foundations of animal ecology, established graduate programmes, and (in the United States) eatablished or transformed wildlife research agencies. In Australia, Ratcliffe was a lone contract employee for CSIR, and there were no local training programmes in the field; in New Zealand there were even less. That began to change just after the war. Both countries and Canada established agencies for wildlife research – the Canadian Wildlife Service, the Wildlife Survey Section of CSIRO, and the Ecology Division of New Zealand's DSIR. The Canadians had the simplest task. The Wildlife Office of the Dominion Park Service dated from 1912, had legal responsibilities for wildlife (in the national parks and the Northwest Territories), and administrators had been planning an independent office for a decade.[16] In Australia and New Zealand the agencies were a response to rising pest populations during the war and had to be built from the ground up.

Both agencies relied on people trained abroad. Francis Ratcliffe, in charge of the Wildlife Survey Section, was from Britain, and Kazimierz Wodzicki, head of DSIR's Ecology Division, was a Polish biologist and former diplomat. In hiring others, they looked primarily to Britain, an established practice – and not only in science. The only new note here was the influence of the Fulbright exchange programme, established as part of American Cold War strategy. They also sent promising people to Britain for graduate study, another time-honoured practice which was supported by scholarships designed to strengthen the ties of home and empire. Their most common destination was the academic group that Charles Elton had established at Oxford in 1932, the Bureau of Animal Population. Wodzicki seems to have been partial to Oxford, and it was, for Ratcliffe, literally the 'old school tie'.[17] It was also one of the few institutions to offer graduate study in animal ecology, and its focus on applied research fitted the agencies' needs. It gave short courses on various subjects and techniques, attracting people from fields ranging from agriculture to public health. In addition, it was becoming an international gathering place, drawing not only students from the Commonwealth countries, but government scientists and faculty and students from the best academic programmes in the United States.[18] Hiring from Oxford and sending students there was the cheapest and fastest way for the new groups to learn about the latest work and to obtain people with up-to-date training.

As in other areas of life, the introduction of new ideas into wildlife policy caused friction. This was most obvious in New Zealand, where ecology challenged the basis of both deer and rabbit control programmes. After World War II, New Zealand decided to apply science against rabbits, but it did so in the context of a reorganized and intensive campaign of conventional control. Parliament established a new national board in 1947, provided for field operations to be carried out by professionals, and abolished the commercial rabbit industry, a step debated in Australia but never taken. Farming in New Zealand provided a high enough return per acre to pay for control; aerial operations were cheap; and stockowner interests were strong enough to override all others – and so the Rabbit Destruction Council began to cover the country with poison baits. It prosecuted its campaign with enthusiasm and rigour. It frowned on rabbits as pets and it 'is said that the Council wished to prohibit the sale of stuffed rabbits as cuddly toys'. This may have been apocryphal, as the subject was 'never mentioned in its Annual Reports', but the board was 'determined to get the last bunny'.[19]

'Work on rabbits was the bread and butter of Ecology Division in the 1950s.'[20] Much of it was on formulating and distributing poisons, but division scientists, sceptical of eradication and the effectiveness of poisoning, also pursued other lines of research. Wisely, they left it to an American visitor, Walter Howard, a biologist from the University of California, Davis, who spent 1958–59 in New Zealand as a Fulbright fellow, to raise these awkward questions. As an outsider he could be frank, and he apparently was.[21] He did not change the programme, but he did give it a reason to look at new questions. In the mid-1960s, for example, the division convinced the Wairarapa Rabbit Control Board to suspend all operations on a one-thousand hectare tract for three years to test the assumption that piles of poisoned rabbits meant progress. At the end of that time the rabbit population was unchanged. Indeed, it might even have decreased. The board was 'extremely reluctant to accept the implications of this trial'.[22] Eventually, however, research and continued experience convinced the authorities that only so much could be done. In 1988 ecologists recommended a general cut-back in control measures on the grounds that rabbits had largely been brought under control. There were dissenters, including the Wairarapa Rabbit Control Board, but it was done, and now rabbit populations, say the scientists, have been 'stabilized at low densities almost everywhere, primarily by natural processes, and pasture production is threatened only in the semi-arid tussocks grasslands of the South Island'.[23] There is no talk of eradication.

Like rabbits, deer had begun as a favoured species and become a pest. From the early twentieth century, the government progressively removed protection as the growing herds threatened farmers' crops and plantations of imported pine. Following the 'deer menace' conference of 1930, all protection was removed and the authorities began an extermination campaign. It was one of the largest organized efforts ever made to eradicate a large mammal. Hired hunters were to sweep through the country, clearing out each district; when they were finished, the deer would be gone.[24] The programme rested on two beliefs: that deer were eating up the forests; and that they could be eliminated by killing individual deer. The first idea was common in New Zealand and Australia. Independently of specifically Darwinian beliefs, people held, almost as an article of faith, that the more 'advanced' European species would 'inevitably' sweep aside the 'backward' native ones, which were 'less fit' in the 'struggle for life'. Its champion here was the eminent New Zealand botanist Leonard Cockayne, whose pronouncements shaped the programme. New Zealand forests, he argued in 1928, had evolved 'unexposed to the attacks of grazing and browsing animals, the moa (*Dinornis*) excepted', and so had no defence against them. Now deer were menacing the woods and, since they were still being protected:

> long after their baneful influence in destroying great national wealth was plainly manifest, it is surely time to protest emphatically ... [T]hese priceless forests of ours are in imminent danger of being transformed into debris-fields and waste ground, and the water which they controlled became the master, pouring down the naked slopes after each rainstorm, bearing with it heavy loads of stone, gravel and clay to bury the fertile arable lands below and occasion floods in the rivers. Is the protection of deer and the like to be permitted to lead to such disaster?[25]

Not if Cockayne could help it. The other idea – that killing deer would reduce the population – was common throughout the settler lands. Grounded in 'common sense' and a simple view of natural processes and connections among species, it had been the backbone of pest control since Anglo-Saxon days and of hunting regulations since the mid-nineteenth century.

Eradication already seemed a doubtful prospect by the time World War II broke out, but the campaign resumed after the war. The first challenge to the policy came in 1949, when an American–New Zealand team did a study of wapiti in Fiordland National Park. Against the conventional wisdom, the study concluded that there was no need to eradicate the animals; a moderate population would not affect the forest. Olaus Murie, now New Zealand's first Fulbright Fellow and a member of the group, made his own report to the Minister of Internal Affairs, suggesting that the extermination policy be reconsidered and fieldwork be done to discover the impact that deer were having on the various forests.[26] Shortly thereafter, the Department of Internal Affairs hired an American biologist, Thane Riney. New Zealanders, Ross Galbreath said, had known that wildlife populations could be managed by 'methods based on ecological principles', but they had had no one to do it. Estimates 'which had been made of deer numbers, of their impact on plant growth or erosion, or of the effectiveness of deer destruction operations had all been based on qualitative, subjective observations – in other words, on guesswork'. For twenty years there had been no real study 'of how effective [the deer destruction campaign] actually was in terms of reducing either deer numbers or the forest damage or erosion the deer were said to be causing'.[27] Now they had someone who 'came from the fountainhead – or near it'. Riney had just finished his master's degree at the University of California, Berkeley, under the direction of A. Starker Leopold, the eldest son of game management's most prominent founder, Aldo Leopold.

Riney was something new in New Zealand science. Graeme Caughley, who began as a government deer shooter and went on to become a wildlife biologist (for CSIRO Australia, among others), said that Riney's attitude was 'entirely alien to Major Yerex' (head of the extermination programme) and that he got 'in hot water with the Department [of Internal Affairs] because he had scant respect for holy writ' (the department's assumptions about deer and forests) and 'set about examining these assumptions as they were hypotheses'. In doing this, he used techniques from ecology and game management, supplemented by common sense and scepticism.[28] To check the age and composition of the herds, he asked the deerstalkers for information about the animals that they shot. That got him a sample, if not a representative one. It also gave the deerstalkers a feeling of being involved and consulted, always useful in projects that depended upon public goodwill. He checked for change in population density by counting droppings along transects (lines drawn through an area), a technique that game managers had refined during the 1930s. He went beyond biology to policy, however, using maps to show that there was little overlap between the areas of high deer density, where the Wildlife Branch concentrated its shooters, and the lands that the Forest Service and the Soil Conservation and Rivers Control Council considered to be in danger of eroding. He cast doubt on the effectiveness of hunting by posting observers with high-powered telescopes to watch hunters in the valley below. On a day when a government shooter saw four deer and shot one, the observer noted twenty-three occasions on which a deer had simply moved out of the hunter's way.[29]

These conclusions were not entirely welcome. It was bad enough for Riney to suggest that the programme had an impossible goal; worse was his conclusion that it was built on a false premise. Deer, he said, were changing the composition of the forests, but they were not eating them to the ground. Cockayne's nightmare of denuded slopes and cities swept into the sea was just that – a bad dream. Worst of all, on one watershed he showed that it was sheep, not deer, that were causing erosion. He perhaps complicated matters by being American at a time when anti-American sentiments ran high. Caughley said that although Riney 'mastered the intricate ritual of the New Zealand tea ceremony he floundered on submerged reefs in the poorly charted channels of the New Zealand psyche … '. He left in 1958, having 'single-handedly advanced New Zealand wildlife research by about twenty-five years'.[30]

Galbreath's history of the New Zealand Wildlife Service provided a more measured contrast, focusing on the differences between Riney and Yerex. The latter had a New Zealander's 'feeling for the native bush and the importance of retaining it unspoiled; Riney, as an American and a scientist, took a more detached view ... '. The two had 'very different views of the place and role of a scientist in the organization'. Riney sought to answer fundamental questions about animal populations 'and for validation of his work ... looked to the wider scientific community. Yerex, on the other hand, expected deference to his authority and loyalty to the organization.' He believed that the scientist's role was 'to investigate specified problems and to provide information to the administration, which would then set priorities and decide upon action'.[31] Scientists, to use the jargon, should be 'on tap, not on top'. Beyond Galbreath's analysis there was a more fundamental division. Yerex, though he might not have phrased it in quite this way, acted as if nature was to be grasped directly and understood by common sense. Certainly, he organized and conducted the extermination campaign as a simple military operation. Riney was not only sceptical of the programme, he did not see observation as the key to understanding. He built up his picture of the forests and the deer from quantitative observations taken within the context of theory.

Events did more than research to reshape policy. In the early 1970s rising prices for deer meat and by-products (used in traditional Oriental medicine), coupled with mobile coolers, scope-sighted rifles, and aeroplanes, produced a slaughter somewhat like that on the North American Great Plains a century before, when railroads and rifles had made it profitable to kill buffalo for their hides. When a four-man team, shooting deer from a small helicopter, could take a hundred out of the high country in a day, the government's foot-bound hunters became redundant, even an embarrassment. As tallies fell and prices rose, farmers agitated for the legalization of deer farming. Eradication was abandoned and deer became a livestock crop. Control operations still continue in some areas, but they are spot controls for what is now defined as a local, not a national, problem.[32]

Conclusion

The existence of what I have termed the Anglo world, and the conditions within it, constituted a key element in these events. That these are a set of countries with a common language and core culture has allowed an easy circulation of expertise and experts, while wide disparities in population and resources have encouraged the less populous to build their scientific institutions around that fact. Ecology did travel from the centre to the periphery, and birds of passage or expatriates did carry it, as convention has it, but other elements do not follow that story. It came not as a developed theory or even a discipline, but as a perspective that shaped research. Institutions of academic science were not its primary transmitters or receivers, and it is hard to see the agencies as steps towards independent national scientific communities – if by that we mean something like those in Europe or the United States. CSIR, which supported Ratcliffe, and New Zealand's counterpart, the DSIR, developed out of efforts to mobilize science for defence during World War I and, after the war, did the same for economic development. They acted to transmit information and ideas, not develop them, and relied on external institutions as a matter of course.[33] This was a rational response to conditions, but it only loosely fits the classical picture of the diffusion of science. Our models need to take more account of Anglo connections as a continuing reality, not as a stage in development.

The public's assimilation of ecology, in the environmental movement, has followed this same pattern. People accepted its perspective on nature long before it provided evidence for its conclusions, and they did so for the same reason that the interwar managers did. Its vision of nature as a set of complex interconnected systems made sense of the problems of pollution and the destruction of nature which they saw around them, and it promised to guide a search for solutions. Like the scientists, the public learned of it through existing channels of communication. Local knowledge and history, however, decisively affected the ensuing national debates. Only the culturally tone-deaf would mistake an Australian discussion of wilderness for one in the United States, or believe that New Zealanders meant the same as Americans did by the term 'national park'. Constitutional divisions of power and Australian history, which left control of natural resources and the land with the states, make the organization of the Australian environmental movement different from the American one, where the federal government is the locus of power in these areas. The size of the Australian scientific community means that it also has different internal arrangements and ties to the public.[34]

Both public and scientific assimilation of ecology are part of a larger, continuing process. For two centuries, since the rise of natural history, Anglo settlers have been seeking to understand their new lands with the aid of their culture's organized knowledge of nature. The events sketched here are part of the latest chapter, still being written, in their quests. It promises to be a vital one. Ecology's perspective challenges ideas inherited from Europe and rooted in settler culture by the experience of expansion. It speaks of limits, and shows how near they are. The settler societies have stressed the inexhaustibility of the land and built their dreams on the chance of wealth for all. It sees people as part of the system of the land and speaks of us as citizens of a biological community. The settlers have spoken most often of conquest. One textbook of environmental law called environmentalism 'one of the great social movements of the century'.[35] It is, and it has only just begun its transformation of our ideas and actions. The study of ecology and empire will be a fruitful topic for many years to come.

Notes

1. Alfred Crosby, *Ecological Imperialism* (Cambridge: 1986); Geoffrey C. Bolton, *Britain's Legacy Overseas* (London: 1973).
2. Robert M. McIntosh, *The Background of Ecology* (New York: 1985), provides the best introduction to the development of the field.
3. George Basalla, 'The Spread of Western Science', *Science*, 156 (1967), pp. 611–22, is the standard. Critiques are in R. W. Home, *Australian Science in the Making* (Cambridge: 1988); Nathan Reingold and Marc Rothenberg (eds), *Scientific Colonialism* (Washington, DC: 1987).
4. Charles Elton, *Animal Ecology* (1927; London: 1966), p. xi.
5. Stoddard to Leopold, 5 December 1931 (Correspondence, Aldo Leopold Papers, Department of Wildlife Ecology, University of Wisconsin Archives, Madison, Wisconsin).
6. See Ronald Tobey, *Saving the Prairies* (Berkeley: 1981), pp. 57–75.
7. See Aldo Leopold, 'Thinking Like a Mountain', in *A Sand County Almanac* (1948; New York: 1966), pp. 137–41.
8. Herbert L. Stoddard, *The Bobwhite Quail* (New York: 1931).
9. Francis Ratcliffe, *Flying Fox and Drifting Sand* (London: 1938), p. vii.
10. Ratcliffe, *Flying Fox*, p. 5.
11. Ratcliffe, *Flying Fox*, p. 317.
12. Ratcliffe, *Flying Fox*, p. 322.
13. Ratcliffe, *Flying Fox*, p. 331.
14. Francis Ratcliffe, *The Rabbit Problem* (Melbourne: 1951), pp. 3–4.
15. Thomas R. Dunlap, *Saving America's Wildlife* (Princeton, NJ: 1988), pp. 74–6.
16. Boxes 30, 38, U 300, Record Group 84, Records of the Dominion Park Service, Public Archives Canada, Ottawa.
17. Author's interview with John Flux, DSIR ecologist, June 1990.
18. Peter Crowcroft, *Elton's Ecologists* (Chicago: 1991), pp. 156–64, 86–9.
19. John A. Gibb and J. Morgan Williams, 'The Rabbit in New Zealand', in Harry V. Thompson and Carolyn King (eds), *The European Rabbit* (Oxford: 1994), p. 166.
20. John Gibb, 'A Brief Review of Rabbit Research in Ecology Division since the 1950s', unpublished paper prepared for Ecology Division, DSIR, March 1983 (Library, Ecology Division, DSIR, Lower Hutt, New Zealand).
21. Interview with Flux, 1990.
22. Gibb, 'Brief review'; Gibb and Williams, 'Rabbit in New Zealand', p. 166.
23. Gibb and Williams, 'Rabbit in New Zealand', p. 163.
24. Graeme Caughley, *The Deer Wars* (Auckland: 1983).
25. Leonard Cockayne, *Monograph on the New Zealand Beech Forests*, vol 2: *The Forests from the Practical and Economic Standpoints* (Wellington: 1928), pp. 2, 10, 11.
26. Ross Galbreath, *Working for Wildlife* (Wellington: 1993), pp. 68–9.
27. Galbreath, *Working for Wildlife*, p. 70.
28. Caughley, *Deer Wars*, p. 70.
29. Galbreath, *Working for Wildlife*, pp. 70–1.
30. Caughley, *Deer Wars*, p. 71. I am indebted to Caughley, who died in 1993, for two interviews about his research and the state of the field in English-speaking countries.
31. Galbreath, *Working for Wildlife*, p. 72.
32. Caughley, *Deer Wars*, pp. 107–17.
33. C. B. Schedvin, *Shaping Science and Industry* (Sydney: 1987).
34. See Libby Robin, 'Of Desert and Watershed: The Rise of Ecological Consciousness in Victoria, Australia', in Michael Shortland (ed.), *Science and Nature* (Oxford: 1993), p. 142.
35. G. M. Bates, *Environmental Law in Australia* (Sydney: 1992), p. 16.

CHAPTER 6

Vets, viruses and environmentalism at the Cape

William Beinart

Introduction

In the early decades of the twentieth century, South African government offi-cials in the Departments of Agriculture and Native Affairs became deeply perturbed about the state of the natural pastures, or veld, which covered the great bulk of the country's surface area.[1] Their concerns mirrored those in a number of settler countries where farm stock were bred on a vast scale over a long period.[2] With respect to the white-owned farms, South African officials advocated the modernization of pastoral farming practices. They wanted an end to 'kraaling' (returning animals each night to the farmstead or byre) and 'trekking' (transhumance), and the introduction of more systematic fencing, dispersed water provision, and rotation of grazing camps. The state sponsored and funded such improvements, as well as the extermination of predators, especially jackals, which hampered them.[3]

Similar prescriptions also had a major impact when they were transferred to the areas occupied by Africans. They helped to feed a growing conviction amongst white administrators and policy-makers in the region that a far-reach-ing transformation of settlement and land use was essential.[4] The imperatives of pastoral reform became intertwined with concerns about population growth, urban migration, social control and the maintenance of segregation. These were invoked during attempts to force African people who lived in scattered settle-ments with unfenced pastures into villages and to change their systems of stock-keeping by introducing fenced camps.

Links between the ideas of environmental conservation, improvement and development were not new, nor were they specific to South Africa. We may understand more about their logic and power by exploring a deeper archaeology of colonial scientific thinking. Richard Grove, amongst others, has begun to excavate these strata in his work on imperial conservationism. In nineteenth-century India, he argues, environmental concerns were formulated primarily by British medical officers addressing problems of desiccation, drought and famine.[5] Their training, often in Scottish universities, provided them with the intellectual interests and botanical knowledge necessary to appreciate the extent of environmental change, notably deforestation. In the Cape during the 1860s, conservationist ideas found expression in the reports of John Croumbie Brown, the Colonial Botanist, who sounded dire warnings about the damage being done to colonial production and natural resources by fire and overgrazing.[6]

The critique of Cape pastoral practices was taken up by the first two Colonial Veterinary Surgeons, both of whom had been working in Scotland. As in the case of the doctors in India, they linked health, in this case of animals, and environmental issues. Their ideas had important implications for patterns of stock-keeping in the colony. Consideration of veterinary prescriptions at the Cape also reveals fascinating examples of conflict between the priorities of colonial scientists and settler farmers about how to cope with a harsh and unpredictable environment which evoke later clashes of perception between colonial officials and African people.[7]

However, evidence from the late nineteenth-century Cape should also caution us against drawing simple lines of division between scientific specialisms on the one hand, and 'local knowledge' on the other. Settler farmers were unified neither socially nor in their approach to stock management. Many of the most ardent improvers, who helped to set the modernizing political and technical agenda of the colonial state, came from their ranks. Both they and the vets were aware of innovation in Britain and in other settler colonies, particularly Australia. While the vets' approach was framed by metropolitan scientific knowledge, they were, in a 'moving metropolis', part of an interactive process of research and policy formation which both drew on and impacted on local knowledge.[8] Moreover, scientific and technical ideas were by no means stable. The first two government vets, William Catton Branford and Duncan Hutcheon, tended towards different interpretations of disease which reflected the rapidly changing science of the time.[9]

Branford and environmental explanations of disease

After the introduction of wool-bearing merino sheep to the Cape in the early nineteenth century, wool had become the major export by the 1830s. Even when it was overtaken by diamonds in the 1870s, wool remained by far the most significant agricultural export commodity. More than 8 million woolled sheep were recorded in the 1865 census; 100,000 angora goats provided a profitable sideline in mohair. But by the 1870s, after a few decades of rapid growth, perceptions of the health of the pastoral economy became more uncertain and a ceiling in wool production appeared to have been reached. From 1872 to 1876 prices were unusually high. Yet most farmers could not take advantage of these favourable conditions by increasing production. Wool exports remained static, even declined, despite the fact that the area under sheep was still expanding and Free State supplies were increasing.

Although the total number of woolled sheep in the colony had grown modestly between 1865 and 1875 to nearly 10 million, the major increases were registered in areas where merinos were relatively recent introductions, in the arid north and the wetter grasslands of the recently colonized eastern border districts, abutting the Transkeian African chiefdoms. By contrast, sharp falls were recorded in some of the best-established sheep-farming districts in the eastern Cape, including areas settled by progressive and successful English-speaking farmers such as Albany and Fort Beaufort. In the semi-arid Karoo midlands, heart of sheep territory, numbers were static.

Concerns about 'decadence' in pastoral farming prompted the new Responsible colonial government to appoint both its first Colonial Veterinary Surgeon and a commission to enquire into stock diseases in 1876.[10] William Branford, who qualified in London in 1857 and worked at the Royal Veterinary College, Edinburgh, served as colonial vet until 1879.[11] His reports, together with that of the commission, provide a startling record of a pastoral economy under pressure. In his view, the flocks and herds were seething with disease. Self-confident in his role as the outside expert, he lost few opportunities to offer remedies.

European settler farmers, amongst whom Branford largely worked, had long categorized the diseases that they encountered by Dutch names which described the animals' symptoms or the parts affected: for example, *brandziekte* (burning disease or scab), *lamziekte* (lame sickness), *stijfziekte* (stiffness); *longsiekte* (lungsickness or pleuropneumonia), *meltziekte* (spleensickness or anthrax) and *hartwater* (water on the heart). The names reflected a concept of medicine that was widespread amongst rural Afrikaners, that diseases were definable by their symptoms and, to some extent, curable by treating those symptoms. Farmers' medical prescriptions, while framed by European origins, also incorporated indigenous plants and knowledge.[12] Some of these names for diseases had been adopted and anglicized by the English-speaking farmers.

Branford saw one of his tasks as imposing an Enlightenment order on this naming system, trying to identify which diseases were known to Victorian veterinary science. Scab, which followed woolled sheep around the world, and lungsickness, a major cattle killer, were well-known in Europe and had already been identified at the Cape. Nevertheless, some farmers held that the *brandziekte* which they experienced was specific to the Colony or their locality. Branford wanted 'to remove an impression which has taken possession of the minds of some of the colonists, viz., that the diseases to which cattle and sheep are liable in the Cape Colony are widely different to diseases which are met with in ... Europe.'[13] If he could establish this, he felt, the relevance of advances in European veterinary science and the application of scientific solutions would be plain.

By no means all stockowners saw science as helpful. Legislation had been passed for the isolation of infected cattle during the Cape's lungsickness epidemic in the 1860s. The lines were being drawn in the battle over scab control. Most Australian colonies (increasingly seen as a model by Cape officials) and progressive sheep farmers had successfully confronted scab by stringent legislation governing stock movements, together with dressing (washing) or dipping with chemicals. In 1874 the Cape government passed a Scab Act, which, although not compulsory, recommended similar controls. Farmers resented interference in their cherished beliefs about stock-keeping, especially the necessity for seasonal transhumance. By emphasizing the uniqueness of Cape animal diseases, they could convince themselves that Branford was another interfering British official who did not understand their country.

Branford was not isolated, however. He found strong support for his ideas among the interlocking agrarian and political élite who had steered through the Scab Act. They had a broad commitment to pastoral reform, but, as investors in increasingly expensive stud animals, they also had a particular interest in reducing the risks of disease. Some felt that they would never be able to compete with Australia until they controlled diseases and used paddocks.[14] After the wool boom of the early 1870s, prices gradually tailed off in a market dominated by expanding Australian production, and this inhibited investment by Cape producers. The mid-Victorian British state had taken on added responsibilities for animal as well as human health, notably after the 1865 cattle plague (rinderpest), a turning-point for the status of veterinary science.[15] Leading vets such as James Simonds had argued powerfully and, after the loss of about half a million cattle, successfully, for stringent controls over imports and stock movements. Branford, who participated in the implementation of these controls, brought with him a belief in the role of the state: diseases had to be addressed at a colony-wide level or the threat of reinfection would remain.

Although Branford held strong scientific opinions, he studied and practised before the breakthroughs by Pasteur and Koch had been widely accepted. Following the dominant ideas of his time, he tended to seek explanations for disease primarily in environmental conditions rather than in microscopic bacteria and viruses.[16] This did not preclude recognition of the role of contagion and infection in specific diseases such as rinderpest and lungsickness. But after conducting a number of post-mortems on sheep, he became convinced that internal parasites, rather than contagious disease or even scab, were the major problem at the Cape. Many of the sheep that Branford examined were infested with worms or, in the wetter districts, fluke. Sheep in Fort Beaufort district, where mortality rates had been amongst the highest, nurtured three or four different kinds of worms; one was 'literally eaten up alive by parasites'.[17] Perhaps most destructive was the wireworm (*Strongylus contortus*), found in huge numbers in the fourth stomach of sheep; it had been identified scientifically by Simonds in 1858.[18]

Farmers called the disease that had killed so many of their sheep 'heartwater'. It was so named because the sheep suffering from the condition, after developing dropsical swellings under the throat, were found to have large amounts of fluid in their hearts. 'Upon opening the sheep', George Judd of Queenstown explained to the Stock Diseases Commission, 'there is little or no blood in the body … [T]he heart when opened discharges a great quantity of clear fluid, and a small quantity of very dark clotted blood.'[19] According to another witness, if you hung a sheep that had died from heartwater, 'nearly a bucketful of water runs out'.[20] An Afrikaner, Grobbelaar, noted that he had lost all but 600 of his flock of 4,000 to heartwater.

Some farmers also gave the name 'heartwater' to slightly different symptoms. In these cases the colour of the liquid was yellow; it did not drip when exposed to the atmosphere but formed a jelly. They noted that this heartwater, also called 'gallsickness', could affect apparently healthy sheep that 'would go out of the kraal fat and well, and come home in the evening with their heads drooping and bodies soft and flabby, next morning they would all be dead'.[21] Although he recognized that there might be a variety of causes for different types of heartwater, Branford felt that the evidence pointed to fluke and worms as the major scourges. Farmers had not generally appreciated the severity of wireworm, he argued, because even those who conducted post-mortems were not all able to navigate as far as its stronghold in the small fourth stomach.

The primacy attributed to fluke and worms reflected the environmental explanations of disease in which Branford and his colleagues had the greatest confidence. The life cycle of the liver fluke was well known to scientists: dependent on marshy land, it was rarely found in the dry Karoo but had spread in better-watered grassland districts. Infestation could be directly linked to stock drinking in vleis and was worse in rainy seasons. Branford argued that it could be controlled by fencing vleis and marshes to control access, or by draining them. Draining was attractive to farmers not only because fencing was expensive, but because vleis were often the richest land for crops. Heavier and more efficient American ploughs which came into use at that time facilitated their cultivation.

The approach to worms was different. Branford and the commissioners thought that the prevalence of worms was linked to overstocking, the deterioration of pastures and the system of kraaling. They sought to demonstrate degradation by showing that the average weight of sheep had fallen by about ten pounds (five kilograms) since the early 1860s.[22] Overstocking, Branford argued, meant that sheep had to 'work harder for a living' and suffered in the competition for food.[23] The term 'overstocking', so often cited in the twentieth century as an evil relating to African peasants, was explicitly used in this nineteenth-century context with reference to settlers. Similar comments had long been made about Australian pastures.[24] By the 1870s, some farmers worked with a clear notion of how much stock a particular type of veld could 'carry' and were able to specify this with regard to their own farms and districts. Overstocking was firmly linked to soil erosion: 'the gradual washing out of the *sluits* and gullies which not only cut up and intersect the *veldt* in all directions, but are evidence of the denudation of the soil which has been very much accelerated since the occupation of the land by sheep'.[25]

Some believed that overstocking had also led to less obvious changes: a reduction in the amount of grass, as well as the growth and spread of 'obnoxious and poisonous herbs' which animals avoided.[26] Branford expanded on the idea of uneven or selective grazing patterns: as the plants favoured by sheep were more heavily grazed, the seeds of those they disliked could easily find more space and spread. This process, he argued, not only impaired nutrition, but had a specific relationship to worms and health. It was widely understood, with good reason, that animals required an intake of saline substances for their health. Afrikaner farmers generally had some knowledge of the 'brak' plants or *brak-veld* that served this function and tried to ensure that these formed part of the diet of their sheep. Their abundance in the Karoo and their relative scarcity in some of the grassland districts to the east was one reason why the semi-arid zones were seen as particularly good for small stock. Overstocking, it seemed, had diminished the cover of 'salt and saline bushes ... which helped to check the great increase of these parasites'.[27] In Australia, the commissioners noted, saltbush (*Atriplex* species) had been strongly associated with the control of sheep parasites.[28]

Some farmers did administer doses of salt as a preventative and cure for internal parasites. For Branford, salt supplements were an essential part of systematic stockbreeding, not an optional ingredient to be given when the sheep showed a craving. Nor were supplements an alternative to the restoration of healthy veld and a pastoral system which allowed valuable plants to recover. Branford saw the most deeply rooted problem as 'the abominably injurious system of KRAALING'. Kraaling was necessitated by the threats of theft and predators. In a rural world still largely innocent of fences, animals left to run overnight could also stray. Shepherds, predominantly black, were used to the system, which had partly been adapted from indigenous stock-rearing practices.

Branford drew together the arguments against kraaling in powerful language which suggested an almost moral imperative to this element of improvement. At the core of his argument was the health risk. Kraaling facilitated the spread of disease of all kinds by bringing animals into huddled contact every night amidst 'gaseous matters, and miasma' which affected blood and tissues. Scab acari and worms could transfer themselves, and contagious diseases found a perfect environment. Many farmers exacerbated the infectious milieu by allowing the collection of 'faecal and excrementitious material' in the kraals, so that sheep would literally stand knee-deep in dung, especially when it was wet, for 'a night's misery'. If they lay down, manure and mud coated their fleeces. 'How in the name of goodness can health be expected in animals exposed to such influences', he expostulated.[29]

By concentrating manure near the farmstead, kraaling deprived the veld of fertilizer. The seeds of bushes favoured by sheep would also be deposited in the kraals rather than in pastures. Daily treks to and from pastures weakened the sheep and destroyed tracts of pasture around the farmhouse and on the droving routes. The commissioners targeted that evocative word 'tramping' to describe the process which they thought had made choice farms 'almost useless'.[30] Their solution was fenced paddocks, where the animals could stay out at night. British experiences of enclosure and rotation were an important reference point. Australian graziers were also fencing rapidly.[31] In fenced pad-docks, sheep would walk less, graze more evenly, deposit manure and seeds, grow fatter and yield more wool. If sections of pasture were rested, seeds would germinate and restore both the balance of the veld and the health of animals. The saving of labour costs in shepherding, a major argument for fencing in Australia, was not so strongly emphasized at this time in the Cape.

For Branford, nothing epitomized the 'decadence of sheep farming' more than the practice of kraaling. 'To those whose farms are "sheep-sick"', he argued, 'I would say, sub-divide, under-stock, and without delay sow.' Healthy flocks required careful management and improved veld. Branford's forceful approach clearly made enemies, and his term of office ended towards the close of 1879 in a bitter wrangle with the government over payment.[32] Nevertheless, a new Contagious Diseases Act, for which he had pressed, was passed in 1881, and the experimental farm which he had advocated so that stock diseases in the eastern Cape could be examined materialized quickly at Leeuwfontein near Fort Beaufort in 1880. It had been considered one of the best sheep farms in the area, but was almost cleared by heartwater. His advice on fluke, according to one Afrikaner parliamentarian, was 'so valuable, that it was followed by all the farmers, and in a very short time the fluke disappeared'.[33] Even in those cases where he was wrong, the environmental thinking that was espoused by him and the commissioners left a strong legacy.

Hutcheon and germ theory

Branford was replaced in 1880 by Duncan Hutcheon (b.1842) who qualified in Edinburgh in 1871 and subsequently worked in Scotland. Hutcheon was imme-diately landed in a punishing programme of disease control and, for most of the 1880s, he had to work alone as the sole qualified government specialist. Like Branford, he saw his annual reports as a means of publicizing his work and ideas. They contain detailed and fluent accounts of his experiences, revealing a strong curiosity for epidemiological material in this unique southern laboratory.

Hutcheon did not necessarily differ from Branford in his broad recommen-dations for the transformation of livestock farming. He could also be critical of farmers. In his first attempt to implement the new Contagious Diseases Act to control an epizootic of lungsickness amongst valuable angora goats in 1881, he ordered the slaughter of infected animals. He was alarmed by the rash of farmers' meetings 'characterizing [the Act] as being only a futile wrestling against Providence, and likely to prove ruinous to the country'.[34] Hutcheon was, however, more cautious about advocating radical changes in farming and tended to be more diplomatic than Branford. He survived in office to become Chief Veterinary Surgeon over a much expanded service in the 1890s, despite the politicization of scientific knowledge, which intensified in direct relation to the state's determination to act upon it.[35]

Hutcheon's caution arose partly because he doubted some of Branford's diagnoses and prescriptions. He had clearly been swept up in the exciting discoveries of germ theory which were gaining widespread acceptance in the veterinary and medical world; comparative medicine was at the cutting edge of scientific advance in the 1870s.[36] It is important not to exaggerate the discontinuities, but whereas Branford's first instincts were to concentrate on worms, pastures and the muck and miasma of the sheep kraal, Hutcheon first looked for invisible 'viruses' or 'poisons'. This led him to question environmental explanations and to suggest that an expanding knowledge of contagious diseases was most likely to fill the gaps in understanding.[37] The diseases with which he was initially confronted, especially lungsickness, probably reinforced his position. In addition to immediate slaughter of infected animals, he strongly advocated inoculation.[38] His approach was, however, also evident in his responses to more mysterious ailments, such as heartwater amongst sheep and redwater in cattle.

Hutcheon's first headquarters were at Leeuwfontein experimental farm, where he began a series of tests to determine the causes of heartwater. Starting from Branford's conviction that it was linked to internal parasites and a lack of saline ingredients in the veld, he divided a newly introduced flock into two and tried to feed one with supplements of salt, saltpetre and sulphur. This cleared the worms, but made little impact on the death rate from heartwater. The symptoms of worm infestation – slow wasting, with external swellings – also seemed to differ from those of heartwater. As coastal farmers had argued during the 1877 enquiry, this heartwater, where the liquid was more yellow, affected apparently healthy sheep. It became clear to Hutcheon during 1881 that there were two distinct sets of symptoms in the diseases called heartwater. One was caused by internal parasites; but the type that he was investigating, the major killer in the coastal districts, was related neither to worms nor malnutrition.

A number of farmers in coastal districts had expressed scepticism about the link between disease and degradation. Denuded farms, they argued, could be free of heartwater, whereas sheep died on others where the pasturage seemed excellent and cattle thrived.[39] The disease seemed to be worse amongst sheep which had grazed in kloofs; some thought that a shade-loving plant was the cause.[40] Hutcheon accepted that other features of the microclimate of kloofs, such as higher levels of moisture, might be a factor, but he rejected the explanation that heartwater was caused by a plant.

Although both Branford and Hutcheon were aware that ticks could carry disease, Hutcheon, like Branford, passed over suggestions that heartwater had followed the spread of the bont (tortoiseshell) tick, on the grounds that the disease was absent from some tick-infested areas. He came to the conclusion that heartwater was a 'blood disease ... caused by some specific virus which gains access to the circulation and sets up some morbid changes within it'.[41] It was sufficiently similar to anthrax, he thought, for him – excited by Pasteur's experiments in this field – to import vaccine from France.[42] The long sea journey and hot overland trip to Fort Beaufort may not have left the vaccine in top condition, but, after two failures at inoculation, he concluded that heartwater was likely to be distinct. He felt sure, however, that the disease was also 'specific', with a definite incubation period and regular pattern of occurrence.

Hutcheon was now disturbed that the disease was 'always ... associated in men's minds with overstocking' – a legacy of Branford's view.[43] He was especially sensitive about public criticism of the state of Leeuwfontein, where he had now been in charge since 1880, and was keen to emphasize that the farm's pastures and stock were in excellent condition except for this 'peculiar disease'. By 1884 he was sure that heartwater had 'no connection whatever with exhaustion of the soil', but resulted from some more 'subtle and intangible ... poison'.[44] Hutcheon developed a more comprehensive criticism of easy assumptions about overstocking, arguing that the coast lands were probably healthier when they had carried larger numbers of sheep which deposited rich dung. Overgrazing was certainly possible, especially in semi-arid Karoo districts, but it was also widely felt, as Brown had reported twenty years before, that heavy grazing in grassland districts could help to convert 'sour veld' into more nutritious 'sweet veld'. While he was an enthusiastic advocate for fencing and paddocks, he did not see them as a cure for heartwater.

In his investigations into heartwater, Hutcheon shifted away from environmental explanations; he took a similar approach to the outbreak of redwater among cattle in 1882–3. He did not know the cause of the disease. But his interpretation of its spread helped him to develop a highly critical approach to trekking or transhumance. As in the case of scab eradication, his emphasis was on the need for veterinary controls which had significant implications for pastoral practices and environmental management.

A recent confrontation over trekking had taken place in the arid western and northern districts of the colony. There was a regular pattern of movement here between the higher, colder mountain areas and the state-owned *trekvelden* in the arid Karoo which were used for winter grazing. From the 1860s, some of the *trekvelden* were leased or alienated on a permanent basis. Officials argued that the government should derive income from the large tracts of Crown land which had been available for minimal payments. Control over trekking would stimulate investment in water provision, thereby diminishing the need to migrate. By the 1870s, sheep numbers in these arid districts had risen sharply; the costs for farmers who had depended on seasonal movement were seen to be offset by an overall increase in production, colonial wealth and government revenues. The critique of transhumance dwelt on classical Victorian motifs of improvement, investment, and the benefits of settled agriculture; to these were added concerns about the role of trekking in tramping veld and spreading disease.

In 1882, redwater, which had been prevalent in the 1870s, reappeared in a virulent form around Kingwilliamstown and East London. Hutcheon was convinced that the disease 'was spread mostly by means of transport'.[45] He attempted to impose the 1881 Contagious Diseases Act, prohibiting the movement of animals in and out of a designated 'infected' zone. He also advocated that farmers along transport corridors should use fencing to prevent access by passing animals. Hutcheon's measures had serious implications for the patterns of trekking and commercial transport in the border districts and he was again faced with an outcry. As a result, the government appointed the Redwater Commission in 1882.[46] Economic arguments lay at the heart of opposition to the restrictions; the only alternative to trekking was rapid and costly investment in water sources and fodder for winter. William Snyman, farmer and chairman of the Komgha Afrikaner Bond, captured the essence of the problem: 'people must travel in this country and must outspan'.[47]

Confusion about the nature of redwater made it difficult for Hutcheon to win the argument. The disease was thought to have come from the United States, where it was more widely known as 'Texas fever'. Endemic in Texas, it had spread to other southern states and even to the Midwest, partly because of the huge traffic in cattle between the southern prairies and Chicago meat markets.[48] Redwater had been associated in the United States with ticks, but veterinary opinion remained sceptical about this link. In both countries, it was most virulent in hotter, wetter districts and seasons.

Branford, in keeping with his general views, initially identified redwater with a disease prevalent in Scotland caused by 'impoverished pastures, heathy moors, and woody districts'.[49] Some witnesses to the commission in 1883 also doubted that the disease was infectious: 'I think that Redwater is just as unintelligible as Heartwater in sheep', William Snyman maintained in an argument that must have unsettled Hutcheon; 'if it is the same as Heartwater, it will be sure to come in the course of time.'[50] Although it was clearly more prevalent on the coast, Hutcheon did not think, as did most farmers, that it was necessarily restricted to this zone. He was sure that it was a predictable, probably infectious disease, either transmitted directly between animals or through food, water or excretion. His attempts to develop a vaccine were unsuccessful, although farmers did have some success in 'salting' cattle. Only after the disease spread inland in two further epizootics was he able to enforce systematic controls on the movement of stock.

In 1892 Hutcheon reported on the experiments in Texas which showed conclusively that redwater was carried by ticks. Newly aware of the significance of ticks as vectors, experiments at the Cape demonstrated that heartwater in sheep was also a tick-borne disease, as had been suggested twenty years earlier by a few Albany farmers. Recognition of the role of ticks transformed treatments. Dipping of sheep for scab, compulsory in some districts after 1886 and throughout the Cape after 1894, was initially only partially effective against heartwater. It took a good deal more experimentation before an effective dip was found.[51] Compulsory dipping of cattle for East Coast fever, a more devastating tick-borne disease, finally brought redwater under control in the early decades of the twentieth century.[52]

Environmental interpretations and environmental effects

In a context where professional knowledge was very limited, the first two Colonial Veterinary Surgeons at the Cape tried to develop diagnoses, preventative measures and treatments for an extraordinary range of animal diseases. Both learnt a great deal from farmers and, more indirectly, from African farm workers. Both, however, tried to make this information systematic in the light of their British professional training and in a way which was sometimes at odds with local knowledge. Although they experienced opposition, Branford and Hutcheon nevertheless found support for their views within the government and amongst progressive farmers.

Branford, who held office when germ theories of disease were still in their infancy, was particularly sensitive to environmental explanations and solutions. He drew on and elaborated emerging imperial and local discourses about environmental degradation; there were many subsequent echoes of the formulations that he and the Stock Diseases Commission developed. While he was clearly wrong about the causes of particular diseases, Branford's environmental prescriptions helped to cement particular ideas about improvement. His more holistic, environmental approach might now be seen as appropriate again, although it was also highly interventionist.

Science, however, was not static. Hutcheon's strong conviction about the validity of germ theory led him to look more discretely at each disease. In making his career at the Cape and as head of an expanding veterinary service, he was also able to observe and investigate diseases over a longer period. While he did not rule out environmental causes, he was more cautious about the links between disease and degradation. His approach suggested different methods of treatment, especially experiments with inoculation. In fact, the major breakthroughs at the Cape at the turn of the century came as much from an expanded understanding of tick-borne diseases as of viruses; compulsory scab legislation in 1894 was also critical. Hutcheon's efforts were increasingly absorbed in combating external parasites, scab acari and ticks, which ultimately entailed universal dipping and control over stock movements. This was a formidable administrative and political task in a dispersed, racially divided and socially fractured colonial society.

Despite shifting scientific understandings, Hutcheon's overall perception of the imperatives of improvement, including fencing, fodder and more careful pasture management, did not differ greatly from Branford's.[53] He was less enamoured with Karoo bushes as cure-alls. But his experiences led to an increasing impatience with transhumance and kraaling. Investigations into *lamziekte* convinced him of the need for supplements in the winter months.[54] He also made clear links between good pasture and good health. On his first visit to the scenic, mountainous district of Barkly East, he was struck that 'the vegetation everywhere was green and luxuriant ... [T]here was no poverty, no disease, amongst stock except a little scab in sheep and lungsickness in cattle.'[55]

But Hutcheon's pattern of thinking and his appreciation of the impact of viruses invisible to the naked eye signalled an important shift in veterinary approaches. Although invisible miasmas did have a role in Branford's explanatory system, the environment and its influences were predominantly those of the visible landscape of earth, animals, insects, worms and plants. Degradation was judged by fire-ravaged or denuded veld. His healing message included an aesthetic element, revealed in views of verdant pastures, trees which would add 'beauty' as well as wealth and hedgerows: 'Treeless fields and bare wire or stone-wall fencing', he noted, 'present a cheerless, uninviting appearance to the eye.'[56]

Hutcheon was more interested in laboratory work. In his long battle to convince farmers that the scab acari was a tiny but living insect, he even took his microscope on the road, showing specimens religiously taken from local sheep to curious but suspicious Afrikaners at country hotels and farms.[57] For Hutcheon, the microscope had the power of explanation – though he by no means convinced all of his audiences. Hutcheon had a far keener sense of contagious agents, their specific identity and patterns of behaviour, filling the void that had been occupied for an earlier generation by bad air and miasma. Understanding of viruses and bacteria was not necessarily inimical to environmental interpretations. These microscopic agents gradually acquired a physical reality for denizens of the laboratory; patterns of reproduction and behaviour could be observed or deduced in a way not totally dissimilar to the more visible natural world. Focus on the microscopic world, its dynamism and the multiplicity of its actors, could potentially enlarge concepts of nature and ecology, as it has done in the longer term.

In some ways, therefore, medical and veterinary advances opened the way for increasingly complex environmental explanations, even when germ theory began to predominate. A synthesis of approaches could – for example, in Theiler's early twentieth-century investigations of *stijfziekte* and *lamziekte* – help to disaggregate diseases which had been lumped together because of superficially similar symptoms.[58] The expanding field of tropical medicine, which had necessarily to explore the behaviour of vectors such as mosquitoes, tsetse flies and ticks, required a multifaceted approach sensitive to environmental influences.[59] In other contexts, such as tuberculosis epidemiology and the discovery of malnutrition in African colonies in the 1920s, significant strands within medical thinking continued to take environmental factors into account.[60]

Nevertheless, it may be correct to surmise that veterinary approaches in South Africa in the early decades of the twentieth century, as in the case of medical science more generally, increasingly divorced explanations of disease from the broader social and physical environment. T. D. Hall, one of the first to attempt a systematic history of South African pastures, noted in 1942 that the 'dazzling achievements of our great Veterinary Research Institute have, until recently, blinded almost everyone to the fact that the millions of cattle and sheep and other animals they have saved have contributed a most substantial share to the loss of South Africa's soils'.[61] Yet the early vets were concerned with precisely this relationship. They saw improved pasturage as essential in increasing numbers and restoring the health of the flocks and herds. And they thought that well-managed flocks could contribute to the maintenance of healthy veld.

It is ironic that the treatments that they pioneered facilitated the processes of overstocking and soil erosion which Branford in particular had so decried. Whereas many farmers found disease a major barrier in building up stock numbers in the 1870s and 1880s, this constraint increasingly fell away in the early decades of the twentieth century. Controls over disease, together with vermin extermination campaigns, fencing and water provision, were key elements in the rapid increase of small-stock numbers from less than 30 million in 1904 to nearly 60 million (including 44 million woolled sheep) by 1930. After a long period of relative stagnation, South African wool production expanded rapidly in these years. By contrast, sheep numbers in Australia, where the initial controls over disease and natural resources had been achieved more rapidly, seem to have stagnated, as the costs of more intense exploitation and rabbits were counted.[62]

While these advances increased the immediate returns from wool, however, they preceded the full range of improvements which the early vets had advocated. The costs of this disjuncture became more obvious to officials in the 1920s and 1930s, so that the transformation of pastoral practices remained a major issue. Depression and drought in the early 1930s, when perhaps 14 million small stock were lost, hammered the message home. In the longer term, many of the early veterinary prescriptions were followed through, although with far more success on white-owned private farms than in the African-occupied communal lands.

Notes

I would like to acknowledge the support of a British Academy/Leverhulme Trust Senior Research Fellowship.

1. Union of South Africa, *Final Report of the Drought Investigation Commission*, UG 49-1923; Union of South Africa, *Report of the Native Economic Commission, 1930-32*, UG 22-1932.
2. G. V. Jacks and R. O. Whyte, *The Rape of the Earth: A World Survey of Soil Erosion* (London: 1939), is the most striking international survey at the time.
3. W. Beinart, 'The Night of the Jackal: Sheep, Pastures and Predators in South Africa', published in French as 'La Nuit du chacal: moutons, pâturages et prédateurs en Afrique du Sud de 1900 à 1930', *Revue Française d'Histoire d'Outre-mer*, 80/298 (1993), pp. 105-29.
4. W. Beinart, 'Soil Erosion, Conservationism and Ideas about Development: A Southern African Exploration, 1900-1960', *Journal of Southern African Studies*, 11/1 (1984), pp. 52-83; see also *Journal of Southern African Studies*, 15/2 (1989).
5. Richard Grove, *Green Imperialism* (Cambridge: 1995).
6. Richard Grove, Chapter 9, below; see also 'Early Themes in African Conservation: The Cape in the Nineteenth Century', in David Anderson and Richard Grove (eds), *Conservation in*

Africa: People, Policies and Practices (Cambridge: 1987), pp. 21–39, and 'Scottish Missionaries, Evangelical Discourses and the Origins of Conservation Thinking in Southern Africa 1820–1900', *Journal of Southern African Studies*, 15/2 (1989), pp. 163–87.

7. Compare Roy M. MacLeod and M. Lewis (eds), *Disease, Medicine and Empire* (London: 1988), and M. Vaughan, *Curing their Ills* (Cambridge: 1991).

8. Roy M. MacLeod, 'On Visiting the "Moving Metropolis"', *Historical Records of Australian Science*, 5/1 (1982), pp. 1–16.

9. Compare Michael Worboys, 'Manson, Ross and Colonial Medical Policy', in MacLeod and Lewis (eds), *Disease, Medicine and Empire*, pp. 21–37.

10. Cape of Good Hope, *Report of the Commission Appointed to Inquire into and Report upon Diseases in Cattle and Sheep in this Colony*, G.3-1877.

11. *The Veterinarian*, 38 (1865), p. 915, and 42 (1869), p. 624; estate papers at the Cape Archives, MOOC 6/9/296, 1516.

12. L. Pappe, *Florae Capensis Medicae Prodromus*, 2nd edn (Cape Town: 1857).

13. *Report of Colonial Veterinary Surgeon [CVS] for the Years 1878–79*, G.54-1879, p. 1.

14. The relative absence of predators was seen as 'a very great advantage Australia possesses over the Cape': *Agricultural Journal of the Cape of Good Hope*, 9 May [1889], p. 176; C. J. Watermeyer, 'Notes by a Cape Farmer on a Recent Australian Tour'.

15. Iain Pattison, *The British Veterinary Profession 1791–1948* (London: 1983).

16. Compare the anthrax story in Australia: see Jan Todd, *Colonial Technology: Science and the Transfer of Innovation to Australia* (Cambridge: 1995), pp. 33–107.

17. *Report of the CVS on Sheep and Cattle Diseases in the Colony*, G.8-1877, p. 17.

18. 'Diseases of Animals: The Discovery of Wire-worm', *Agricultural Journal of the Cape of Good Hope*, 27 July [1893], p. 284, taken from the *Journal of the Royal Agricultural Society*, 24 (1863).

19. *Report of the Commission ... into ... Diseases in Cattle and Sheep*, p. 4.

20. *Report of the Commission ... into ... Diseases in Cattle and Sheep*, p. 42.

21. Evidence of W. Ayliff, MLA, *Report of the Commission ... into ... Diseases in Cattle and Sheep*, p. 120.

22. *Report of the Commission ... into ... Diseases in Cattle and Sheep*, p. 12.

23. *Report of the CVS on Sheep and Cattle Diseases*, pp. 10–11.

24. The 1877 commission specifically drew on a range of contemporary Australian sources: on the Australian experience, see Geoffrey Bolton, *Spoils and Spoilers* (1981; Sydney: 1992); Elinor G. K. Melville, *A Plague of Sheep* (Cambridge: 1994).

25. *Report of the Commission ... into ... Diseases in Cattle and Sheep*, p. xiii.

26. John Shaw, 'On the Changes Going on in the Vegetation of South Africa Through the Introduction of the Merino Sheep', *Journal of the Linnaean Society, Botanical*, 14 (1875), pp. 202-8.

27. *Report of the Commission ... into ... Diseases in Cattle and Sheep*, p. xv.

28. *Second Annual Report of the Secretary of Agriculture* (Melbourne: 1874), esp. pp. 150–5.

29. Quotes from *Report of the CVS on Sheep and Cattle Diseases* , pp. 10–11.

30. *Report of the Commission ... into ... Diseases in Cattle and Sheep*, p. xxi.

31. W. H. L. Ranken, *The Dominion of Australia* (London: 1874), pp. 8off.

32. Branford to Colonial Secretary, 4 July 1878 (Cape Archives, Colonial Secretary's Papers, CO 4197, B52; CO 4203, B44, B46, B50, B52; CO 4208, B34).

33. Evidence of J. A. Burgher, MLC, *Report of the Select Committee Appointed by the Legislative Council to Consider and Report upon the Appointment of a Minister of Agriculture*, C.2-1882, p. 42.

34. *Report of the CVS for the Period from March 16th 1881 to February 28th 1882*, G.25-1882, p. 42.

35. Conflicts in the 1890s centred around scab and rinderpest control: *Report of the Scab Disease Commission, 1892–94*, G.1-1894; C. van Onselen, 'Reactions to Rinderpest in Southern Africa,

1896–1897', *Journal of African History*, 13/3 (1972), pp. 473–88; P. Phoofolo, 'Epidemics and Revolutions', *Past and Present*, 138 (1993), pp. 112–43.

36. Lise Wilkinson, *Animals and Disease: An Introduction to the History of Comparative Medicine* (Cambridge: 1992).

37. *Report of the CVS from 1881 to 1882*, p. 42.

38. *Report of the CVS for the Year 1880*, G.52-1881.

39. *Report of the CVS from 1881 to 1882*, p. 64.

40. The link between disease and poisonous plants was also made in Australia: see Todd, *Colonial Technology*, pp. 33–45.

41. *Report of the CVS from 1881 to 1882*, p. 66.

42. *Supplementary Report by the CVS, 18 May 1882*, A.88-1882; cf. Todd, *Colonial Technology*, pp. 66–7.

43. *Report of the CVS for the Year 1883*, G.45-1884, p. 78.

44. *Report of the CVS for the Year 1884*, G.31-1885, p. 44.

45. *Report of the CVS for the Year 1882*, G.64-1883, p. 62.

46. *Commission of Inquiry into the Disease among Cattle Known as Redwater* G.85-1883.

47. *Redwater Commission*, pp. 182–3.

48. J. F. Smithcors, *The American Veterinary Profession* (Ames, Ia: 1963), pp. 436ff.; William Cronon, *Nature's Metropolis: Chicago and the Great West* (New York: 1991).

49. CVS to Colonial Secretary, 9 January 1877 (Cape Archives, Colonial Secretary's Papers, CO 4193, B4).

50. *Redwater Commission*, pp. 182–3.

51. *Agricultural Journal of the Cape of Good Hope*, 9 June [1898], pp. 696ff.; Smithcors, *American Veterinary Profession*, pp. 449–50.

52. Paul F. Cranefield, *Science and Empire: East Coast Fever in Rhodesia and the Transvaal* (Cambridge: 1991).

53. *Report of the CVS for 1883*, p. 79.

54. *Report of the CVS for 1884*.

55. *Report of the CVS for 1883*, p. 4.

56. *Report of the CVS on Sheep and Cattle Diseases* , p. 12.

57. *Report of the CVS for the Year 1886*, G.14-1887.

58. A. Theiler, 'Stiff-Sickness or *Stijfziekte* in Cattle', and 'Facts and Theories about *Stijfziekte* and *Lamziekte*', *Agricultural Journal of the Union of South Africa*, 1/1 (1911), pp. 10–21, and 4/1 (1912), pp. 1–56.

59. Worboys, 'Manson, Ross'; Vaughan, *Curing their Ills*, p. 34.

60. Randall M. Packard, *White Plague, Black Labour: Tuberculosis and the Political Economy of Health and Disease in South Africa* (Pietermaritzburg: 1990); Michael Worboys, 'The Discovery of Colonial Malnutrition', in D. Arnold (ed.), *Imperial Medicine and Indigenous Societies* (Manchester: 1988), pp. 208–25.

61. Thomas D. Hall, *Our Veld: A Major National Problem* (Johannesburg: 1942), p. 6; William Beinart, 'Environmental Destruction in Southern Africa: Soil Erosion, Animals and Pastures over the Longer Term', in Thackwray Driver and Graham Chapman (eds), *Timescales and Environmental Change* (London: 1996).

62. Bolton, *Spoils and Spoilers*; Neil Barr and John Cary, *Greening a Brown Land* (Melbourne: 1992).

CHAPTER 7

Enterprise and dependency: water management in Australia

J. M. Powell

Too many 'centre' and 'periphery' designs are stale promotions for an overused resort.[1] Conventional analyses of Australia's imperial connection undervalue important reciprocities, impressive evidence of local initiative, and productive interchange with other settler societies. Furthermore, notwithstanding the seemingly ubiquitous influence of investment flows and the bonding relationships of high science within the British sphere, Australia's colonists also relied on their own vernacular brands of science and technology and on a negotiation of diverse relationships with the wider world. Perhaps all too obviously – given its accepted status in standard histories of the driest of the inhabited continents – 'water management' in its many guises has been fundamental in the Australian settlement experience.[2] Water management narratives, highlighting environmental appraisal, resource development and community responses to ecological change over the period 1788–1950, shed new light on the imperial connection.

Empiricists and synthesizers, 1788–1880

This first section draws on the contexts of ecological change in south-eastern Australia before the 1880s. It looks at pastoralism, mining operations, irrigation planning, dryland farming frontiers and, briefly, at the management of new urban centres. The evidence prompts a questioning of casual assumptions about the emergence of derivative cultures at the edges of empire.

Australia's 'squatting' frontiers are relatively well known for their contributions to economic, political and social change. In these respects, settlers are probably too lightly cast as the vanguards, beneficiaries and pawns of imperial expansion. The squatters' highly independent explorations delivered the first reliable inventories of extraordinarily prodigious expanses of territory, including elementary topographical maps illustrating the distribution (and sometimes the painful absence) of major and minor river systems. At the level of the individual run or station, insecure tenures discouraged comprehensive interventions in local ecosystems. Natural grasslands and savannas were generally preferred to forest and woodland situations, but large property sizes seldom afforded anything resembling drought-proofing, and the financing of small dams, wells and augmented water-holes was very common. On the broader

Adapting to Australia: the necessity for a water management perspective on the 'Australian experience'. From the *Bulletin*, 10 December 1903.

'A drought-resisting stock'

front, official compilations of regular livestock returns eventually yielded practical regionalizations which were the summations of precocious measures of 'proven' carrying capacity encompassing vast areas. The rapid grasp of a stunning new spatial scale was a signal achievement in itself. This early environmental reconnaissance could not be ignored during the implementation of revolutionary American-style policies for the introduction of small-scale 'yeoman' or homestead farms in the 1860s.[3]

Contemporary appreciations of surface water resources, early indications of environmental degradation, and apparent variations in rainfall reliability were subsumed and consolidated in these important records.[4] Australian settlers knew from the earliest times that they were committed to a type of *empirical testing* in which water resources featured very prominently. British backers and clients depended on the success of an intimate assessment by early squatters and farmers that they could neither share nor fully comprehend.

The history of mining adaptations amplifies the point. First, Victoria's early gold miners were hugely reliant on niggardly water supplies, and opportunistic individuals were soon profiting from the sale of water 'rights'. British law guaranteed neither efficiency nor equity in dry climate undertakings, and new legislation was passed in the 1860s to assert – or, rather, reassert – community rights.[5] The role of miners from California in devising ingenious cross-country

water transfers is generally acknowledged and, in so far as they were already
familiar with American furores over water rights, it is reasonable to suppose
that their lively presence was also noticeable in these Victorian disputations.
Water rights took on greater significance during the introduction of govern-
ment-sponsored irrigation schemes in south-eastern Australia and on the South
Island of New Zealand.[6] Secondly, Victorian gold-field communities helped to
present the case for the historic 'integrated' or multi-purpose Coliban water
supply project to cater for irrigation, domestic needs, mining operations and
waste disposal.[7] It was no modern behemoth, but warrants citing for its administrative and legal precedents and for its value as a staging post for peripatetic
colonial engineers. The original design was conceived by an Irish engineer who
had trained in England. It attempted to answer colonial arguments about timber
clearances in vulnerable watersheds and local rehearsals of the ancient tree–
rainfall equation. Its execution was marred or contaminated by local
pettifogging, and it was shot through with colonial–imperial tensions. Although
it is thoroughly representative of its time and place, it is also a useful historical marker for twentieth-century efforts in integrated water management
planning.

DRIVEN TO IT.

Quantity and quality: some
very strong aversions to Melbourne's early water supply.
From the Melbourne *Punch*,
24 February 1881. Connec-
tions with questions of
public health suggested the
adoption of a 'closed catchment' policy and should be
seen as an anticipation of
holistic, environmentally
sensitive river-basin planning
in the twentieth century.

In 1871 the Victorian government commissioned Lieutenant-Colonel Richard H. Sankey of the Corps of Royal Engineers, India, to report on the Coliban.[8] He emphasized the superior claims of technocratic leadership in the production and supervision of what he called 'remunerative public works'. What the colony required was a new agency managed by a specialized 'hydraulic engineer', and India's Public Works Department gave the model. Sankey also proved to be a champion of thorough technical inventories: 'Nothing is easier than to project grand waterworks, nothing probably more difficult than to execute them; as water never makes a mistake.' Yet he praised the preservation of the *public interest* in surface water resources on the mining fields and urged its extension. British India, he explained, already exemplified the idea.[9] In fact, the supporting public interest philosophy had not been considered special or radical in Victoria, where the platform of a Land Tenure Reform League called for the cessation of Crown land sales, repurchase of all alienated lands, abolition of private freeholds, and determinations of rentals as a form of direct tax to replace all indirect taxation. Superficially, the influence of Henry George's 'single tax' seems clear, but the level of activity in Victoria suggests much more than a mere echo of that 'Prophet of San Francisco'. Notable are the pressure for a type of nationalization and the favouring of the leasehold principle. Each objective accrued political and practical urgency in the 1880s. Each would be realized, at least in part, during the introduction of a framework for state-sponsored irrigation which embodied a rich blend of colonial, imperial, North American, Middle Eastern and European approaches. This novel system would be researched and vigorously presented by a future Prime Minister of Australia's young federation, Alfred Deakin.

Victoria's well-advertised economic prospects had attracted increasing numbers of immigrant engineers, and this new talent was recruited to provide some stiffening for the highly politicized 'Selection' programmes aimed at the expansion of yeoman farming frontiers. In 1860, at the onset of a ferment of pioneering which was to occupy two generations of rural colonists, an adventurous prize essay from engineer Frederick Acheson foreshadowed an amalgamation of the lakes of Victoria's Western District into a single regulated storage. He even allowed for the drainage and replenishment of those that were naturally saline.[10] In addition, building on the squatters' rude individualistic adaptations, Acheson suggested ways of reducing the recurring impact of water shortages on the semi-arid north-western plains. He urged the construction of a series of 'branch catchment drains' parallel to the better natural watercourses; further local drains could be cut by plough to lead the normal run-off into these channels, so that the newly constructed system might be boosted during flood times. Half-baked, admittedly, but Acheson also alluded to the prospects for huge storages in the eastern high country to harness the Murray River. His main proposal offered a measure of drought insurance for an area larger than Wales, and the makeshift prescription enjoyed surprising longevity.

Imperial and American precedents influenced a good deal of irrigation

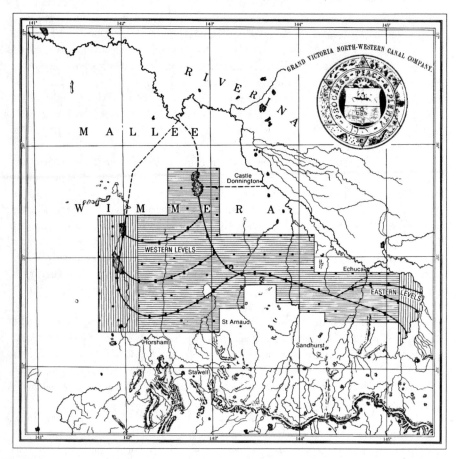

A proposed canal scheme for north-western Victoria. Adapted from the prospectus of
the Grand Victorian North-Western Canal Company (1871). Approximately 130 'irrigation stations', shown here in black, were individually numbered on the map which
accompanied this influential private-enterprise speculation. The striped pattern indicates
the alternating strips of land to be applied for after the fashion of private continental
railroad companies in the USA; intervening areas would remain in government hands.
'Castle Donnington' was later known as 'Swan Hill'. The Mallee-Wimmera region
notionally 'commanded' and otherwise advanced by the scheme encompassed some
74,000 square kilometres – only slightly smaller than the whole of Scotland, from whence
hailed Victorian politician Hugh McColl, a fervent supporter of this 'Grand Canal'.

boosterism. Of the less obscure projections which adopted, refined and magnified Acheson's composition, the most pertinent was conceived by capitalistic
enterprise in 1871. The prospectus of the Grand Victorian North-Western Canal
Company conjured a 300-kilometre main canal with related storages, fed by
the Murray and its tributaries and by steam-pumped groundwater, to be used

principally for irrigation and navigation. Subsidiary canals would be added and, in time, the climate of the inland would be comprehensively ameliorated: thus, 'Queensland and Victoria would shake hands over the well-watered plateaus of Riverina, and the continent have *bona fide* title to its original and prophetic cognomen of New Holland'.[11]

The canal company was brazenly promoted by Benjamin Dods, who sought a partnership with the Victorian government which would emulate American-style 'land grant' subsidies for railway construction: that is, the company would be permitted to sell large amounts of land to defray its expenses, and alternate frontage blocks would be retained by the government. Some Victorians took the whiff of Yankee speculation in their stride, and Hugh McColl, an energetic gold-fields politician who became notorious as Australia's representative specimen of Hydrocephalic Man, served for a time as Dods's secretary. A more intricate regional planning component was vaguely envisaged: for instance, service settlements surrounded by intensive farming units would be located at sixteen-kilometre intervals. Exports along the artificial waterways would 'soon rival the Mississippi trade itself'.

By that time, George Perkins Marsh's *Man and Nature* (1864) had been discovered by Australian intellectuals. Leading scientists and bureaucrats continued to debate the ecological and economic implications of a marked decline in the tree cover, including its relationships with regional precipitation and, therefore, with water supplies. These issues attracted the combined attention of renowned colonial botanist Baron von Mueller and locally prominent members of one of the nineteenth century's premier natural history coteries, the cleric–earth scientist fellowship (which included W. B. Clarke, William Woolls and J. T. Woods). Australian evidence (on overgrazing and the 'murderous practice of ringbarking', for example) was frequently cited in conjunction with gleanings from Marsh, and there was some recognition on the global front of imperial culpabilities in the tropics.[12] Whether perceived as a leading or subsidiary consideration, water management concerns were seldom omitted from bookish forecasts of impending environmental crises.

For the majority of colonists, the ecological argument was better absorbed by hard-edged practical experience and, in this light, events in South Australia put up an unusually robust proposition. By the early 1870s the colonial government had been forced to surrender its Wakefieldian function as supreme arbiter of land quality. Yeomen farmers fanned out over the vast sub-humid north and, for a time, they were rewarded by above-average rainfalls. Pioneers had no way of reckoning that average, unfortunately, and the ingrained reliance on empirical testing strengthened the conviction that 'rain follows the plough'. American echoes still invite close attention, but it is just as important to point to the contribution of experiential learning in the South Australian tradition of regional planning. The returning droughts delivered salutary lessons of 'marginality', and the concessions required for successful adaptation included the development of drought management strategies and a battery of mixed farming regimes capable of exploiting low and variable moisture supplies without sacrificing sustainability – crucial variants of the narrower interpretations of water management and quintessentially, though not uniquely, Australian. These special regional requirements were not generally acknowledged until the spectacular advance–retreat sequence was concluded in the 1880s. It is noted here because of its authentic native source in the mid-1860s. Surveyor-General Goyder's singular effort to produce a practical demarcation to assist decisions on drought relief to pastoralists was purloined by a beleaguered government. The defeated army of empiricists retreated behind this famous demarcation, this 'Goyder's Line', which was then etched into technocratic and vernacular discourse as a hallowed delimitation of 'reliable rainfall'. Popularly and officially, the line continues to document a historic truce between society and nature.[13]

While environmental damage by farming, grazing and mining taxed the best minds and consciences of the day, the under-regulated colonial towns provided equally searching interrogations. In some localities the ecological disruptions appeared all but complete within the span of a single decade. Unmade roads became seasonal quagmires oozing with animal manure, unstable sources of wind-borne filth; pristine rivers received vast quantities of refuse from bustling abattoirs, tanneries and tallow-works, from unsewered houses and the tented fields of new immigrants. By the mid-1850s, the worst areas of the colonial capitals were gothically squalid in their wretched stench and putrefaction. But there was more: periodic outbreaks of dysentery, typhoid, cholera and unspecified fevers (even bubonic plague) frequently ignored the niceties of social geography; because water levelled as often as it liberated, there was a generous range of opinions on its management.

In the gold-rush colony of Victoria the enumerated population increased from 77,000 in 1851 to over half a million a decade later. Presiding over the boom, Melbourne paid the price in wholesale environmental degradation. A dramatic rise in the rates of morbidity and mortality was attributed to deplorable hygiene and sanitation and especially to chronic deficiencies in the supply and quantity of fresh water. The threatening situation captured the attention of immigrant administrators and aspiring technocrats who patronized Victoria's flourishing 'philosophical' societies. Employing these networks, a lively Melbourne press and an impressionable young parliament, the colonial élite fastened on the water challenge. Quite well-informed about British 'sanitary' policy, they admired its aura of respectability, its voguishness, as much as the underlying science and supporting databases. In their own fields, however, Australia's government statisticians were being equally thorough. Certainly, their pursuit of the ruling miasma theory seems no less diligent than that of their British counterparts.[14] Even when it was gradually being replaced by sophisticated germ theory towards the end of the nineteenth century, the highly accessible 'miasmata' proposition continued to stimulate colonial interest in the regulation of many aspects of urban living, including issues of residential density and guarantees of open space, zoning ordinances to quarantine 'noxious trades', the removal of wastes, and efficient water reticulation. At several levels, the changing urban landscapes testified to these growing concerns.

By 1880, however, local and intercolonial interchange had facilitated cross-referrals between administrative milieux, casting doubt on the dependence on imported environmental theories and associated management practices. In particular, leading colonial technocrats ridiculed the *ad hoc* adoption of British and European calibrations of reservoir capacity and drove home the expensive corollary: collections of place-specific precipitation and run-off data should be considered mandatory. Only occasionally was this extrapolated into gung-ho, go-it-alone technological adventuring. More pertinently, it underlined the credentials of colonial operators and legitimated some of their claims for a freer rein, while intensifying local interest in environmentally analogous regions regardless of their imperial or other affiliations. In the interim, Ferdinand von Mueller and Government Astronomer R. L. J. Ellery exposed pervasive timber-cutting biases in the application of half-understood multi-purpose principles to the area supplying Melbourne's Yan Yean reservoir. Supplementing the rising interest in the ecology of human health since the 1850s, these blasts helped to promote an assertively innovative 'closed catchment' policy.[15]

Irrigationists and sanitarians, 1880–1900

This section tightens its focus in order to situate the twin goals of improvement and guardianship in the changing milieu of social and environmental anxieties; it draws again on rural and urban examples of local innovation and colonial and global interchange. Water management concerns were often masked by the Byzantine meanderings of settlement legislation. From the inception of responsible government in the mid-1850s to the end of the nineteenth century, assorted 'land' matters, notably homestead-style laws and related enablements for railways, roads and other infrastructure, packed the agendas of the colonial parliaments. But water management discourse moved up in the pecking order in the 1880s. And its focus was somewhat unusual in so far as it combined urban and rural interests, nominated radical changes in administration and regional economy, and mapped out a route to technology-led progress in a style which was well fortified by an austere summoning of 'authorities'.

Like the forestry debates which had occasionally preceded but more often shadowed them, the discussions on water were shot through with very modern conceits; they warrant more attention.[16] Victorians made the early running in the promotion of irrigation. Conventionally, most of the praise goes to Alfred Deakin, in acclamation of his inspirational chairmanship of the Royal Commission on Water Supply (1884–7) which produced *Irrigation in Western America* (1885). Undoubtedly, this celebrated inquiry upgraded the local exchanges by inserting firsthand accounts of what might now be called 'best international practice'. Deakin made a three-month tour of the United States,

STRIKING THE ROCK.

"THE PRINCIPAL WORK OF THE SESSION WILL BE IRRIGATION." *Hon. Deakin "Moses".*

Alfred Deakin as Moses, the Deliverer. From the Melbourne *Punch*, 3 June 1886. Deakin made good use of American, imperial and European experience, and was advised by local engineers and regional politicians.

where his delegation consulted Colonel Hinton and E. A. Carman of the Department of Agriculture, N. H. Egleston of the Forestry Bureau, and leading engineers in California and Colorado. It noted the opinions of John Wesley Powell, an expert on arid lands, and was alerted to the divisiveness engendered in California by the inheritance of different traditions regarding water rights and to the varying divisions of responsibility between governments, local communities and the irrigators themselves. Its bulging dossier listed modes of supervision and the planning utility of intertwined data on soils, subsoils, slopes, types of crop, precipitation, evaporation, aspect, length of growing season and the myriad factors responsible for determining the 'duty of water', the area of land which a given quantity of water could be expected to irrigate. In addition to Deakin's admirable overview volume, the more specialized reports of engineer J. D. Derry (on engineering practices) and engineer-surveyor Stuart Murray (recommendations for Victoria) were also published, together with Deakin's account of irrigation in Egypt and Italy and a well-received series of newspaper articles.[17]

Much has been made of the departures in the resulting Irrigation Act: the promised 'nationalization' of water use rights, state construction of irrigation facilities, authorization of loans to 'Irrigation Trusts'. The last of these provisions had been prefaced by temporary legislation. There is probably as much evidence for persistence as for change. On nationalization, legal histories cite the precedents of the Spanish Law of Waters (1866) and the Northern India Canal and Drainage Act (1873).[18] Reservations of water frontages had been common in Victoria since the squatting era and had taken on special import during the gold-rush years. Other continuities may be traced. For instance, one of the reasons for convening the commission was the mixed reception accorded to the efforts of Alexander Black and his colleague George Gordon, formerly Chief Engineer of Water Supply, to combat northern droughts in the later 1870s.[19] Supporting a revised version of Acheson's design, they proposed a decentralized system based on a minimum of government interference and administrative recognition of the supremacy of river-basin divisions. Direct and indirect outcomes included reservations of unalienated water frontage strips, the establishment of local Trusts for water supply, and controversy over funding mechanisms and 'experts'.

Accordingly, Deakin was able to craft a number of old and new strands into his eloquent presentations to parliament. His plan to remodel amalgamated Watershed Trusts on a 'natural basis' now seems curiously reminiscent of J. W. Powell's audacious suggestion that the development of the American West might be best assured by the seeding of a kind of technological democracy based on river-basin units.[20] Deakin retreated from this interesting quasi-ecological position. He had larger fish to fry. Nor is it necessary, in any case, to credit Powell alone with the idea: the fundamentals of river-basin planning were hardly novel and Victoria's precedents had ranged from the cautious drafts of Gordon and Black to such cobbled initiatives as the Coliban. Deakin soon veered away sharply to market 'self-help' – albeit cushioned by a combination of state-owned 'national' headworks and locally controlled distributions – while carefully indicating the existence of similar frameworks in Egypt, France, Italy, Spain and British India. Opting for Crown ownership of natural waters appeared to underline the break with English riparian law, but the departure is probably more safely described as 'hybrid'.

Anti-irrigationists charged that there were far more cogent reasons for investment in the lightly settled southern districts, but the government had the numbers and Deakin pressed on to promise the transformation of 'barren wildernesses' with increased population and productivity and safety from nature's blows: so, the 'greater garden' of Victoria would claim its rightful place in the coming federation. Wider influences may be discerned. Deakin was not personally immune to the speculative frenzies which gripped his contemporaries. He served conspicuously in governments which borrowed and invested unwisely.[21] Furthermore, the federation movement itself, which benefited greatly from Deakin's astute guidance, was by no means purely nationalistic:

it was also promoted by a few keen imperialists.[22] Nor was the irrigation initiative removed from this spectrum of convoluted 'centre–periphery' relationships. In the event, it was botched. There were no great storages. The Trust system was routinely manipulated to indulge an existing pattern: the use of supplementary water to shore up extensive grain and grazing operations. The inevitable result was financial collapse. By the end of the century, another drought had mocked the politicians' pretentious announcements. Even the irrigation colony of Mildura, a private-enterprise venture anointed by Deakin and run by Canadian-American entrepreneurs, was foundering.[23]

What, then, of Australia's cities? To the extent that they received any 'gospel of efficiency', it was an unstable compound of British, American and local authorship, laced with laconic agnosticism. Certainly, there was a stubborn reluctance to sink precious capital in mundane pipes and holes in the ground: the need had been palpable for decades and no progress could be made without a change in *mentalité*. 'Progressive' arguments began to force a resolution in the 1880s, when, as often as not, the arresting theme was water *quality,* as compared with the equally worrying but far more 'accessible' questions of quantity.

Colonial statisticians drew disturbing comparisons, based on crises in water quality and quantity, between purportedly Dickensian Britain and the immigrants' havens in Australia. Approaches to urban water management necessarily incorporated evolving conceptualizations of centralized and decentralized forms of city government, architectural and social-psychological perspectives on city planning, environmental health and equity concerns, and the minutiae of borrowing and cost-recovery policies which shaped the transport, gas and electricity 'utilities'.[24] Interdependent representations of urban issues do not constitute a major breakthrough in holistic reasoning. But there is no doubt that 'public health' brought the perils of insidious ecological change into bold definition and that a well-broadcast desire for security, efficiency and amenity reinforced the demand for 'improved' water management.

Preparations for the International Exhibition of 1880 prompted new efforts to sanitize Melbourne, but reformers were still restricted by a plethora of jealous intrametropolitan authorities. American preferences for metropolitan government were reviewed and, while it lasted, London's brand of federated administration commanded respect. An emergent consensus noted that modern sewerage and water supply schemes should proceed in parallel with an overhaul of urban administration and town planning.

Between 1880 and 1889 there were 2,688 recorded deaths from typhoid in Melbourne. A Royal Commission on the Sanitary Condition of the Metropolis (chaired by Professor H. B. Allen, a close associate of Alfred Deakin) produced a devastating report in which telling comparisons were made with other Australian cities and with Britain, Europe and New Zealand. A Public Health Bill was introduced with the aim of placing a supervisory board of experts over the municipal representatives, and although the resulting legislation retained most of the status quo, the thwarted medical experts were temporarily soothed by the appointment of Dr D. A. Gresswell, an experienced British sanitarian, as the board's first medical inspector. The same momentum brought out British sewerage engineer James Mansergh to prepare a civilized and civilizing scheme for Melbourne. Mansergh's lofty advice helped to launch the Melbourne and Metropolitan Board of Works in 1890. This unwieldy 'municipal parliament', initially chaired by Irish immigrant, Edmund G. Fitzgibbon, survived into the twentieth century. Then, for good or evil, it built massively upon its original water management brief to become the dominant planning body for the metropolitan region.[25] These public health issues demonstrated a 'miasmatic' justification for formative environmental policy and bolstered the imperial connection by means of the adaptation of British measures and the enlistment of independent British consultants.

Science, technology and organizational reform, 1900–1950

Histories of technological change commonly give premium status to city communities. As mentioned above, Australia's cities were indeed the hearths and recipients of new ideas, but they were also the beneficiaries and victims of a transformation of the New World that continues to resist caricatured geographical circumscriptions. From the standpoint of environmental themes, other forcing grounds such as the farming, grazing and mining frontiers demand priority. Concentrating on the first half of the twentieth century, this section finds more antecedents of modern water management practice, outlines additional claims for innovation, cites a firmer local grasp of problems of scale and environmental diversity in a huge national territory straddling tropical and temperate country, and illustrates the growing influence of imperial and global pools of administrative, scientific and technological expertise.

British settlers in northern Australia had to devise a range of adjustments to cope with monsoonal and seasonally cyclonic extremes and African- and Indian-scale droughts. Queries about the relationships between climate and health applied to the viability of livestock farming and to the physiological and psychological capacities of the white race; in true Australian fashion, practical experience alternately disputed and complemented scientific and pseudo-scientific hypotheses.[26] Good examples emerged from the production of the inventories required for the management of flood control and reservoir technology. In colonial Brisbane the local managers struggled with a mixture of the 'visiting expert' ploy, empirical observations and straitened finances. Responses to the 1893 floods centred on potential mitigation schemes and pitted ex-army engineer Colonel John Pennycuick, late of India, against J. B. Henderson, the government's overworked hydraulic engineer. Deplorable sanitary conditions in Brisbane and most of Queensland's provincial towns were aggravated by heat and humidity, and although the dangers of primitive waste disposal were only reluctantly conceded, a long-term reliance on reservoir investments was predicted. For urban and irrigation supplies, there was a habitual reliance on imported expertise from the rest of Australia and overseas. There were disappointing returns: for instance, New York's Alan Hazen, imbued with American paranoia about the siltation of expensive storages, rejected local preferences, received an exorbitant fee, and left a legacy of flops and half-successes.[27]

In Queensland, as in every other Australian state, irrigation and dryland farming projects were monitored at the federal and imperial level, the peripatetic expert became the *sine qua non* of investment insurance, and common sense continued to require that Australians be equipped with a modicum of prudent synthesizing to cope with the blizzard of specialist knowledge. Sugar-cane farming commenced in earnest on rough, scaled-down versions of southern American plantations, partly based on the exploitation of imported Pacific-island labourers. It began to assume a more thoroughly Australian character when the industry was made over to small-scale 'yeoman' farmers, whose powerful organizations lobbied state and federal governments at every turn. During the first half of the twentieth century, well-worked partnerships with government brought out international experts with invaluable imperial and American training in the application of irrigation technologies to tropical agriculture. Even so, a good deal of the industry's success was based on local innovations, especially the interpretation and subtle manipulation of coastal groundwater resources. A similar hybrid evolution is evident in the promotion of state-aided irrigation for cotton growing. In academic circles, in contrast with these practical insights and preoccupations, a more sceptical mood emerged. The environmental limits of settlement in general, and the value of Australia's tropical third in particular, shaped the colourful campaign of pioneering geographer Griffith Taylor, but his 'environmental determinism' found skeletons in the citizenship cupboard and his fall from imperial grace was no more damaging than the insults received from bombastic nationalists.[28]

Politicians came and went, and the accumulation of applied environmental knowledge owed more to the long careers of a small number of senior technocrats. J. B. Henderson has been credited with the introduction of regular stream gauging and the establishment of a weather bureau in Queensland, and with Government Geologist R. L. Jack, he also had a large role in the development of groundwater resources in that state.[29] His major water brief was to supply the public roads and stock routes on the seasonally parched western plains, which meant that the geographical scope of his activities was only matched by his counterparts in the British Raj. The persistent appeal for more reliable and accessible water points was occasionally inflated to sketch the revolutionary benefits of 'fodder reserves', great and small, distributed across the inland. Neither 'innovative' nor 'derivative' will describe this elaboration: 'opportunistic' comes closer. The influences of recurrent famines on developmental programmes throughout the British Empire incorporated Irish experiences as well as those in India and elsewhere.[30] In Queensland, as in much of the rest of Australia, the powerful famine argument was subtly adapted to the needs of marginal farming and grazing operations.

However, Henderson's detailed involvement in the management of the Great Artesian Basin undoubtedly built upon practical and theoretical work in other states. Sydney's highly regarded Government Astronomer Henry Chamberlain Russell calculated in 1880 that the surprisingly low average volume of the Darling River was inexplicable without making allowance for accessions to a gigantic and 'probably inexhaustible' groundwater reserve.[31] Initial drilling explorations depended upon North American technology and personnel, and the field attracted both private and government funding. Exploitations for more intensive grazing, together with the detailed landscape textures of the new economic region, mirrored the continuing dialogue between governments and pastoral leaseholders. Australia's groundwater reserves surrendered their secrets very slowly between the late nineteenth century and World War II. The basin is unique in its huge extent and internal differentiation, so local speculations on the diminution of supply, whether folk- or science-based, could not be clarified by overseas experience.[32]

The under-representation of tropical and sub-tropical Australia is a curious weakness in environmental-historical scholarship. Yet it must be admitted that, in water management affairs, the temperate sector in general, and the state of Victoria in particular, maintained an overall lead. Over the turn of the century, drought, depression, the loss of population to the other states and the collapse of the decentralized trusts prefaced a government-led recovery which spawned a monolithic agency, the State Rivers and Water Supply Commission (SRWSC), serving most non-metropolitan districts.[33] It should be stressed that the area to be administered in this, the smallest of the mainland states, was not much smaller than the whole of the United Kingdom: in fact, this SRWSC, notwithstanding its more modest aims, was a historic venture which predated the Tennessee Valley Authority (TVA) by three decades. The prime movers behind its establishment were the Minister of Water Supply, George Swinburne, and the ageing Stuart Murray. It became the forerunner of several statutory corporations which transformed government administration. Murray ensured that the government gained control of those water frontages which had escaped earlier measures, and Swinburne drew on his own practical experience in the gas and electricity sectors to argue for a water rate on available supplies *whether used or not.* Victoria's Closer Settlement Act (1904) had followed a New Zealand precedent by introducing the compulsory purchase and subdivision of large freehold estates, and sophisticated irrigation was seen as an essential component of the accompanying plans for 'intensification'.

Appropriately, Murray was appointed the first chairman of the SRWSC, but he soon made way for an international high-flyer: Elwood Mead, no less, was adroitly head-hunted in 1907 from his position as Chief of the Irrigation and Drainage Investigations Bureau in the United States Department of Agriculture.[34] An evangelical in the American 'Progressive' tradition, Mead appeared to supply the dimension craved by Deakin during the 1880s. Exploiting (rather than advancing) the kind of state socialism pioneered in Australia and New

Elwood Mead, by courtesy
of the Rural Water Com-
mission of Victoria. Mead
was one of a small number
of notable water managers
whose careers embraced
experience in the USA and
parts of the British Empire.

Zealand, Mead persuaded his paymasters to fund pilot irrigation estates, agricultural education, and a sales tour to Europe and the United States. The
SRWSC became influential in Queensland and the Murrumbidgee Irrigation
Area in New South Wales, thanks to Mead's (personally lucrative) consulting
jaunts. After his return to the United States in 1915, he experimented with
state-controlled irrigation colonies, Victorian-style, in California.[35] By the time
of his departure, the resolution of a long battle over the use of the River Murray
for irrigation and transportation was proclaiming the virtues of multi-purpose
planning, federal–state co-operation in natural resource issues, and some ex-
pansions of spatial vision. For instance, South Australia, Victoria, New South
Wales and the federal government backed the formation of a supervisory River
Murray Commission, with instructions to concentrate on the southern borderlands of the giant basin. And the claims of the wider natural unit received an
airing – Hugh McKinney (ex-Indian Civil Service, significantly) had produced
the first reputable map of the Murray–Darling Basin in 1900. That was a small
but necessary step towards the bioregional thinking adopted in the 1980s and
1990s. The purview of the enterprising MDB Commission covers an area as
large as the combined territories of France and Spain.[36]

Interest in the related issues of land and water degradation was another composite of local empiricism and international influences. It is also a good example of the emergence of science's own imperium, encapsulating and transcending the British Empire's impressive circuitry of ideas, careers and institutions. Australian scientists' cherished links to Rothamsted, Oxbridge and the British Association for the Advancement of Science did not dissuade them from sorties into independent fundamental research, nor, indeed, from following the progress of soil science in the United States and Russia.[37] Similarly, the New Deal's responses to the depression and the Dust Bowl found resonances in every Australian state, and the TVA was widely touted as a model for basin-centred comprehensive planning. Soil erosion and salinization had been graphically described by colonial scientists, graziers and farmers, and the ecological bases of those phenomena had become more apparent in the early years of federation. Australia also had its share of well-publicized dust-storm events and, in this regard, even boasted a few near-clones of America's Hugh Hammond Bennett.[38] Its dominating water agencies variously led or encouraged the crusade for regional planning and soil conservation legislation.

Franklin D. Roosevelt's oratorical style struck responsive chords in Australia, where comparable basin regionalizations had been winning approval, but the British 'Catchment Area' concept and New Zealand's 'River Districts' also appealed.[39] Technical and environmental accentuations can be overdone, however. Arguably, extreme American and British proponents of the regional approach were adopting a fundamentalist position that refused to countenance any transgression of the borders 'set' by nature. For contemporary Australia, pragmatists and rationalists of most political persuasions saw enough in the regional idea to take it in their stride, while supporters of the growing Labor party mainly accepted it as a democratizing strategy for expressions of local needs and the efficient delivery of nationally co-ordinated programmes. Wartime conditions brought more urgent reviews of the conduct of all facets of regional planning, partly to improve local self-sufficiency, self-reliance and 'preparedness'. Queensland reorganized its water management administration on regional lines, and Victoria's SRWSC sent H. G. Strom to investigate river control programmes in New Zealand; Strom's report recommended a blend of British and New Zealand regional practice, and more inclusive planning programmes soon embraced the regional idea.[40]

Conclusion: neglected articulations, rediscoveries of place

Too many inferences have been drawn from Australia's increasing integration into the world economy as a child of empire. As Jan Todd relates, more research is needed into a range of undervalued articulations – of imported technology with domestic technologies and economies, in public–private linkages, and between local science and local production technologies.[41] Her detailed case-studies are informed by recent revivals of the 'social construction of knowledge' and a current rediscovery of *place* in the social sciences:

scientific and technical knowledge ... arises within a given context, through the work of people focused on particular problems, situated in particular situations or institutions, surrounded by particular kinds of peers, guided by particular norms of accepted practice. More specifically, there are many factors which go into the design and refinement of a particular technology – theoretical and empirical knowledge, resource and economic constraints, social values and tastes, legal standards and requirements, institutional and organizational structures and assumptions.[42]

Settler societies were not insulated from the time and space convergences which relentlessly undermined locational sovereignties throughout the empire. In the twin fields of resource appraisal and environmental management, however, combinations of novel circumstance, ambition and ingenuity usually negotiated a sizeable measure of independence. It is stimulating indeed to catch glimpses of Ireland in the politics and administration of land settlement; traces of India in water and forest management and of South Africa in land degradation and in scrub and grassland management; rough facsimiles of Britain itself in the debates on the human health implications of changing urban–industrial ecologies; African cadences in environmentally attuned livestock breeding and immunization programmes; and still more Indian borrowings in climatological research. Arguments for the facilitating sense of imperial coherence afforded by the ties of investments, loyalties, and military, academic, cultural and scientific institutions are not necessarily diminished by the fulsome testimony of place-specific engagements with complex environmental themes. The evidence nonetheless demands that greater consideration be given to emphatically local resolutions, as well as to the relevance of autonomous and semi-autonomous communications within the so-called periphery and outside the imperial pool.

Enterprise-dependency equations obviously vary over time and with the issue under investigation. Inescapably, since Australians dealt with the driest of the inhabited continents, 'technology transfer' studies concentrating on water management demand that ecological perspectives be placed in the foreground. By asserting the endemicity of *sui generis* empirical testing and by elevating local innovation and opportunism, even the most compressed sampling of water management narratives makes Todd's welcome revision seem conservative. Acknowledging the importance of distinctly imperial matrices, the water management narrative also tracks major diffusions from other settler societies and especially from the United States, while indicating a crowded record of *in situ* initiatives.

This overview suggests that, whether those initiatives be judged positively or negatively, they are certainly incomprehensible in today's 'environmentalist' reckonings without recourse to assiduous contextualization. The linking thread is a perennial and multifaceted *coming-to-terms* which might be unassailably Australian – were it not for its striking relevance to other settler societies.

Notes

1. The more presumptuous frameworks of orthodox dependency theory may be unrealistically confining; this chapter takes the view that this is emphatically so in the case of Australia. It builds upon a much wider literature, including P. Cochrane, *Industrialization and Dependence* (St Lucia: 1980); S. Cutliffe and Robert Post (eds), *In Context: History and the History of Technology* (London: 1989); Donald Denoon, *Settler Capitalism* (Oxford: 1983); C. Dixon and M. Heffernan (eds), *Colonialism and Development in the Contemporary World* (London: 1991); Barrie Dyster and David Meredith, *Australia in the International Economy* (Melbourne: 1990); R. W. Home (ed.), *Australian Science in the Making* (Cambridge: 1988); Ian Inkster, 'Scientific Enterprise and the Colonial "Model": Observations on Australian Experience in Historical Context', *Social Studies of Science*, 15 (1985), pp. 677-704; Roy M. MacLeod (ed.), *The Commonwealth of Science* (Melbourne: 1988); Jan Todd, 'Science at the Periphery: An Interpretation of Australian Scientific and Technological Dependency and Development Prior to 1914', *Annals of Science* (1993), pp. 33-58.

2. The term is usually applied to freshwater resources; here it is used in its inclusive sense to incorporate water harvesting as well as co-ordinations of allocations between users, provisions for ecological requirements, flood and drought mitigation strategies, and allowances for amenity and recreational values: see J. J. Pigram, *Issues in the Management of Australia's Water Resources* (Melbourne: 1986).

3. J. M. Powell, *The Public Lands of Australia Felix* (Melbourne: 1970), 'Conservation and Resource Management in Australia 1788–1860', in J. M. Powell and M. Williams (eds), *Australian Space Australian Time* (Melbourne: 1975), pp. 18–60, and *Environmental Management in Australia 1788–1914* (Melbourne: 1976).

4. Powell, *Environmental Management*, pp. 31–2.

5. Compare Donald J. Pisani, *To Reclaim a Divided West: Water, Law, and Public Policy, 1848–1902* (Albuquerque: 1992). This specialist contribution criticizes Samuel P. Hays, Donald Worster and others for positing élitist/centralized models and for understating a complex localism and economic individualism which co-existed with expectations of state assistance. Pisani's findings will assist towards an appreciation of the Australian situation.

6. Powell, *Environmental Management*, pp. 37–41; J. M. Powell, *Watering the Garden State: Land, Water and Community in Victoria 1834–1988* (Sydney: 1989), pp. 49–51; T. J. Hearn, 'Riparian Rights and Sludge Channels', *New Zealand Geographer*, 38 (1982), pp. 47–55; J. M. Powell, 'Mining and Land: A Conflict over Use 1858–1953', *New Zealand Law Journal* (August 1983), pp. 235–8.

7. The scheme is located in the Bendigo-Castlemaine district; its early vicissitudes are discussed in Powell, *Watering the Garden State*, pp. 74–84.

8. The disbursement of military engineers lent a certain unity to water management practices around the British Empire. 'Defence' personnel were prominent in resource appraisal and environmental management over a very lengthy period in the Old and New Worlds alike, and many good research prospects are offered: the list includes the production of comparative data on climate and health, several varieties of pioneer mapping and spatial information (down to an early dominance in satellite technology in our own day), extraordinary opportunities for global monitoring incident to the military avocation, and subsidiary lines of environmental import such as quarantine and reservation policies and hazard-reduction or coping strategies.

9. Powell, *Watering the Garden State*, pp. 80–2.

10. A modified version of the prospectus map appears in Powell, *Watering the Garden State*, p. 86.

11. Powell, *Watering the Garden State*, esp. p. 88.

12. The basic message was already being aired in the Australian colonies, but Marsh's book seemed to supply an authoritative ventilation: cf. Powell, *Environmental Management*, pp. 54–64; see also Richard Grove, *Green Imperialism* (Cambridge: 1995), which argues strongly that Marsh's views were anticipated in the tropics.

13. D. W. Meinig, *On the Margins of the Good Earth: The South Australian Wheat Frontier 1869-1884* (1962; Adelaide: 1970). On the broader Wakefieldian context, see Michael Williams, *The Making of the South Australian Landscape* (London: 1974). For related American folk myths, see W. and J. Kollmorgen, 'Landscape Meteorology in the Great Plains Area', *Annals Association of American Geographers*, 63 (1973), pp. 424–41; J. M. Powell, *Mirrors of the New World* (Canberra: 1978), pp. 111–18. The 'official' state history of water management is M. Hammerton, *Water South Australia: A History of the Engineering and Water Supply Department* (Netley: 1986).

14. The best-known of this group are William Henry Archer and Henry Heylyn Hayter (Victoria) and Timothy Coghlan (New South Wales). Environmental perspectives on the miasma theory are offered in Powell, *Mirrors*; there are many American parallels.

15. Patrick O'Shaughnessy, *Melbourne and Metropolitan Board of Works Catchment Policies* (Canberra: 1986).

16. Stephen M. Legg, 'Debating Forestry: An Historical Geography of Forestry Policy in Victoria and South Australia, 1870–1939', Ph.D. thesis (Monash University: 1995).

17. Deakin was born in Collingwood, a Melbourne suburb, to English immigrants. He was Australia's second Prime Minister, 1903–4, and regained the post in 1905–8 and 1909–10. His *Irrigation in Western America* (Melbourne: 1885), the first *Progress Report of the Royal Commission on Water Supply*, was printed separately as a book and did almost as well in America as in Australia. Deakin's *Irrigation in Egypt and Italy* (Melbourne: 1887) drew less on firsthand experience and, like his *Irrigated India* (London: 1893), was too late to influence legislation. His chief colleagues were immigrants: Ellery (Director of the Melbourne Observatory) was born in Surrey; Black (Surveyor-General), McColl (formerly of Britain's Tyne Conservancy Committee) and Murray were all born in Scotland. Devonshire-born Derry had served in the Indian Public Works Department on Punjabi canals before 'retiring' to Victoria; he was associated with water management improvements on the dry Wimmera and Mallee plains.

18. P. N. Davis, 'Australian Irrigation Law and Administration', Doctor of Juridical Science thesis (University of Wisconsin: 1971).

19. Gordon, another Scot, undertook his engineering training in England, spent six years in Holland, with four of those as the Chief Engineer of the Amsterdam Water Company, and chalked up a further ten years as Chief District Engineer of the Madras Irrigation and Canal Company before taking up his appointment, in 1872, as Chief Engineer of Victoria's Board of Land and Works.

20. Compare Donald Worster, *Rivers of Empire: Water, Aridity and the Growth of the American West* (New York: 1985), pp. 132–43.

21. J. A. La Nauze, *Alfred Deakin: A Biography*, vol. 1 (Melbourne: 1965), p. 138.

22. Luke Trainor, *British Imperialism and Australian Nationalism* (Cambridge: 1994).

23. J. A. Alexander, *The Life of George Chaffey* (Melbourne: 1928).

24. Dan Huon Coward, *Out of Sight: Sydney's Environmental History 1851–1981* (Canberra: 1988); Graeme Davison, *The Rise and Fall of Marvellous Melbourne, 1880–1895* (Melbourne: 1978); David Dunstan, *Governing the Metropolis: Melbourne 1850–1901* (Melbourne: 1984); W. Gill, 'Social and Scientific Factors in the Development of Melbourne's Early Water Supply', M.Sc. thesis (Melbourne University: 1981). On Queensland, see E. Barclay, 'Fevers and Stinks: Some Problems of Public Health in the 1870s and 1880s', *Queensland Heritage*, 2 (1971), pp. 3–12; R. Patrick, *A History of Health and Medicine in Queensland 1824–1960* (St Lucia: 1987). On Perth, see S. J. Hunt, *Water: The Abiding Challenge* (Perth: 1980), pp. 5–31.

25. Tony Dingle and Carolyn Rasmussen, *Vital Connections: Melbourne and its Board of Works 1891–1991* (Melbourne: 1991); compare M. Beasley, *The Sweat of their Brows: 100 Years of the Sydney Water Board, 1888–1988* (Sydney: 1988).

26. D. Arnold (ed.), *Imperial Medicine and Indigenous Societies* (Manchester: 1988); Roy M. MacLeod and M. Lewis *Disease, Medicine and Empire* (London: 1988); compare William Beinart, Chapter 6, above.

27. J. M. Powell, *Plains of Promise, Rivers of Destiny* (Brisbane: 1991) pp. 92–3.

28. J. M. Powell, *Griffith Taylor and 'Australia Unlimited'* (St Lucia: 1993).

29. Both Henderson and Jack were trained in Scotland. Henderson's experience in Australia came from the Coliban and Geelong water schemes in Victoria; Jack worked on the Great Artesian Basin in conjunction with exploration for coal, gold and other minerals.

30. For reflections on famine and organizational reform, see S. B. Cook, *Imperial Affinities: Nineteenth Century Analogies and Exchanges between India and Ireland* (New Delhi: 1993).

31. Russell, born in New South Wales, pioneered global approaches to climatology, limnology and meteorology and instituted regular weather forecasting and monitoring, together with systematic river gauging. He also invented self-recording meteorological instruments – a boon for Australia's vast open spaces. For NSW generally, see C. J. Lloyd, *Either Drought or Plenty* (Parramatta: 1988).

32. Powell, *Plains of Promise*.

33. The SRWSC lasted for over eighty years from the early federal era until it was swept away by another wave of reform.

34. Mead was then at odds with his Washington peers, and the Victorian government offered to double his salary. Imbued with a better grasp of social science than many of his technical brethren, he was a pioneer advocate of planned rural settlement. His *Irrigation Institutions* (1903) was well received in Australia, as was *Helping Men to Own Farms* (1920). Cf. J. R. Kluger, *Turning on Water with a Shovel: The Career of Elwood Mead* (Albuquerque: 1992); Powell, *Watering the Garden State*, pp. 150–85.

35. J. M. Powell, 'Elwood Mead and California's State Colonies: An Episode in Australasian–American Contacts, 1915–31', *Journal Royal Australian Historical Society*, 67 (1982), pp. 328–53.

36. J. M. Powell, '*MDB*': The Emergence of Bioregionalism in the Murray–Darling Basin (Canberra: 1993), and *Plains of Promise*.

37. J. M. Powell, *An Historical Geography of Modern Australia: The Restive Fringe* (1988; Cambridge: 1991), especially pp. 121–49.

38. Powell, *Watering the Garden State*, pp. 211–22 ; Powell, '*MDB*', pp. 33–8; see also Thomas R. Dunlap, Chapter 5, above.

39. Michael Roche, *Land and Water: Water and Soil Conservation and Central Government in New Zealand 1941–1988* (Wellington: 1994).

40. Powell, *Watering the Garden State*, pp. 207–22; Powell, *Plains of Promise*, pp. 195–208. The top Queensland manager was Tom Lang, who later took up a senior appointment with Australia's Snowy Mountains Authority.

41. Jan Todd, *Colonial Technology* (Cambridge: 1995).

42. Todd, *Colonial Technology*, pp. 11–12.

PART 3

Nature and Nation

Nationhood and national parks: comparative examples from the post-imperial experience

Jane Carruthers

'... To forge nations out of the chaos of colonial history'.[1]

In terms of international legislation as well as in popular consciousness, a national park was, for more than a century, the most advanced level of protected area; it encapsulated all that was 'good' and 'unselfish' in nature conservation.[2] As is well-documented, the first national parks in the United States crystallized the romantic settler frontier experience which had brought within the fold of a self-conscious new nation some of the earth's most spectacular and monumental scenery. Many works have emphasized the ecological innovation of national parks, but recent research has involved closer and critical scrutiny of its features as a 'social invention'.[3]

In the first half of the twentieth century, the American idea of the national park 'spread to many parts of the world and changed considerably in the course of its travels'.[4] In many British colonies, however the path to landscape protection came by a different route, even though the final product bore the same name. In Australia, where the settler community originally had its roots in urban Britain, the first national park was established on the outskirts of Sydney in 1879. It has been argued that this (Royal) National Park (in fact, the first to have been so called) harks back to the royal parks and open spaces of Britain – for example, Richmond Park.[5] South Africa's experience has been different again, but, like Australia, it owes far more to its imperial connection than to the United States. Because of its spectacular and abundant wildlife, game protection – that 'pleasing British characteristic' as colonial game-warden C. R. S. Pitman expressed it[6] – was the original rationale for the South African national parks. While the United States model lauds national parks because they are 'for the use of the public forever', the concept of what constitutes 'the public' or 'a nation' is not a simple one. Nonetheless, the word still carries with it connotations which were expressed by the *Sydney Morning Herald* in 1888: '"Nation" is a big word to use but there is dignity in it and pride ... though it is rather the symbol of what shall be than the expression of what is ... '.[7] An analysis of protected areas shows that, in the South African case,

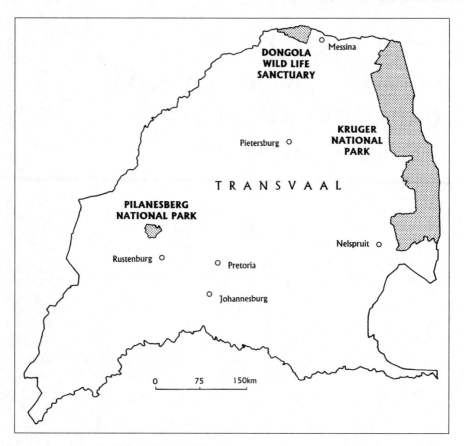

Location of South African National Parks.

but also elsewhere, national parks have been divisive institutions. This chapter
examines three South African case-studies – the flagship Kruger National Park,
the failed Dongola Wild Life Sanctuary and the former apartheid 'homeland'
of Bophuthatswana's Pilanesberg National Park – each of which contains elements that explain how the national park idea is sectional, even ethnic, and has
mitigated against national unity rather than for it.[8]

In modern environmental policy circles, national parks are losing ground. It
could be argued that their disadvantages, particularly in terms of their explicit
repudiation of a human element, are catching up with them. National parks
contain a basic contradiction in that they are saved for people and yet it is a
state duty to protect national park land against people and against change.[9] This
fortress approach has had the consequence that parks are 'islands under siege',[10]
especially in the post-colonial context, because the costs and benefits of these
islands have been borne unequally by different segments of the national population.

In comparison with Australia, where scholarly knowledge of Aboriginal manipulation of the environment is becoming increasingly sophisticated,[11] much remains to be uncovered about pre-colonial use of the wild fauna of southern Africa. From the outset, however, despite the abundance of fauna which they found in the African interior, people from Europe blamed Africans for ruthless extermination. Imperial antagonism for African hunting was based on the ideology that to subsist on game (as Africans did) or to sell it (as Boer settlers did) was 'less civilized' than to kill for amusement. There was also distaste for 'cruel' African methods of hunting. But by the end of the nineteenth century the principal reason had become quite clear: 'the destruction of game by the natives ... enables a large number of natives to live by this means who would otherwise have to maintain themselves by labour'.[12] Wildlife conservation thus played a role in creating a proletariat in the industrializing Transvaal.

The international dynamic of protected areas is strong, initiated at almost the same time by both the European imperial powers and by the United States. The London conference of 1900, which concerned itself with protecting the wildlife of Africa, was the first organized imperial intervention.[13] In his analysis of imperialism and its relationship with the natural world, MacKenzie argues convincingly that it was the imperial hunt which facilitated a protectionist ethos in Britain.[14] In many respects, Africa, while not part of the Old World, was well known to Europe. Its geography – at least in outline – was established, its inhabitants relatively familiar, and its plants and wildlife had been studied for a number of centuries. It had also attracted hunters who thrilled the reading public of Britain with tales of their personal valiance, daring exploits and hairbreadth escapes.

By contrast with Africa, Australia was a new continent in almost all respects, but its unique endemic fauna was initially derided rather than valued by settlers, although naturalists abroad studied it eagerly. To the first whites who saw it, Australia appeared gentle and park-like, but it soon demonstrated that it was a harsh and unyielding environment.[15] Unlike Africa, Australia was not initially attractive to British eyes – perhaps because it was indeed 'implausible', its plants and animals were so very strange and unusual – and early settlers worked hard to convert it into a replica of what they had left behind by introducing plants and animals from Britain.[16] Early colonists appear to have been somewhat ashamed of Australia, 'the home of freakish animals and strange plants ... [where] "trees retained their leaves and shed their bark ... some mammals had pockets, others laid eggs ... even the blackberries were red" '.[17] Moreover, it was apparently difficult to feel spontaneous national pride for a place where the 'fauna was too tame, and the Aborigines insufficiently war-like'.[18]

Just a year after the conference on African wildlife in London in 1900, the Australian colonies federated into a 'nation'. It might have been expected that, retaining closer imperial links than South Africa was to do, Australia would have become involved in the wildlife-saving mania of the penitent butchers in Britain. But it is somewhat ironical that Australia was affected more by the

USA's international conservation thrust of the time than it was by the imperial. Powell shows how ideas of US forestry conservation permeated the new 'white Australia' and how forestry, not wildlife, became the core conservation effort.[19] Griffiths, too, in explaining interest in natural history that came about in Melbourne at this time, describes how a group of nature-lovers acknowledged their debt to Emerson and Thoreau and called their bush hut 'Walden'.[20]

The Kruger National Park

South Africa's first national park was established by the Union Parliament in 1926 by combining two provincial game reserves in the eastern Transvaal which had been founded at the turn of the century during the first wave of modern protectionism: the Sabi (1898) and the Singwitsi (1903). When formal administration of the Sabi Game Reserve had begun after the South African War in 1902, the 3,000 African residents were summarily evicted by the warden, James Stevenson-Hamilton.[21] Before long, however, both he and the Transvaal colonial administration, which was funding the enterprise, realized that labour was needed and that a medieval deer preserve with a lone keeper in command was inappropriate for Africa. Thus the policy of evictions was reversed and remaining tenants within the reserves became liable for labour or rents in cash.

Prisoners supplemented the tenant workforce; illegal immigrants from Mozambique seeking work in the Transvaal gold mines who crossed over game reserve boundaries were arrested by reserve officials as trespassers, which resulted in a fortnight's imprisonment.[22] Having served their sentence – not within prison walls, but on road construction and other projects – prisoners were given a 'pass' which entitled them to work in the Transvaal.[23] Other Africans were in direct game reserve employment, as domestic workers and 'native police' who aided white game-rangers in their duties. Because Africans were not permitted to walk on the public roads, visitors to the national park from the 1920s were not made aware that they were looking at an inhabited landscape. Continuous efforts were made to prevent Africans from enjoying national parks, even as tourists. In 1932, Afrikaner nationalist and historian of the 'volk', National Parks Board member Gustav Preller, was incensed that Indians used the same camp as whites.[24] There was even a serious suggestion in 1949 from J. G. Strijdom, Minister of Lands in the new Nationalist government, that the Kruger National Park be divided like the rest of the country, with one portion for use by whites, the other for 'Bantu'.[25]

When the game reserves were young, African settlement of the region was sparse and those Africans who were dispossessed were able to settle outside the reserve boundaries. As the state increasingly restricted African access to land (particularly by legislation in 1913 and 1936), it became more difficult for Africans to move elsewhere. By the mid-1930s a contest was brewing between the Tsonga Makuleke community between the Levubu and Limpopo Rivers and the national park authorities which coveted the rich riverine vegetation which the Makuleke controlled. For thirty years there was extreme tension between the Makuleke and the Kruger National Park, with the outcome in 1969 – at the height of apartheid – that the community was forcibly removed from land that they had occupied for at least six generations.[26] When moving the Makuleke was first mooted, there was considerable antagonism between the Native Affairs Department, which sought to protect Makuleke interests, and the national park, which sought to curtail them. Even within the National Parks Board there was dissenting opinion, voiced by the secretary, H. J. van Graan: 'Is it wise to take this step …? Frankly, I foresee in this gain of today, if we acquire the Pafuri, the future germ of destruction of the whole Park.'[27] Van Graan's words may well be prophetic, because the Makuleke have now submitted a claim to the far north-east of the Kruger National Park under the Land Restitution Act of 1994.

The conservation ideology which whites propagated in game reserves exacerbated national divisions. Upper- and middle-class European values about hunting for sport and the cruelty of snares and trapping were imposed on Africans whose values were the opposite: killing for sport was wasteful, and snaring was an appropriate sustainable utilization of a natural resource. Harsh sentences were imposed for poaching, which encompassed even the catching of cane rats or tortoises, or chasing a lion from a kill. Incidents of poaching, which were widely reported, entrenched the white perception that Africans living within or near parks could not be trusted.

The success of the Kruger National Park was due to the general adoption by whites of the philosophy that viewing and studying wildlife constituted a legitimate – and financially viable – form of land use and that state land should be allocated for this purpose. This was very much in line with the national park philosophy of the United States. But in 1926 other circumstances in South Africa also coincided to make the national park a reality, including the nascent rise of Afrikaner nationalism, the consolidation of a Voortrekker mythology and the search for a unified white South African national identity. Soon the national park was being overtly exploited to exemplify and inculcate white South African culture.[28]

The cultural importance of the Kruger National Park developed particularly after 1948 with a body of literature emanating from the National Parks Board which not only equated nature protection with whites, but with Afrikaners in particular. This view of a close connection between Afrikaners and national parks owes more to mythology than to history. Part of the myth of the Afrikaner as conservationist must be appreciated in the context of the task of unifying Afrikaans-speakers and liberation from an imperial past.[29]

Generally speaking, white Australian society was more homogeneous and never threatened to split along 'racial' lines. There were other cleavages, however, which affected the structure of protected area reservation and management. In particular, as Powell puts it, 'patriotism in Australia … [was] … enunciated at two levels, imperial and parochial, with little thought for nationalism's intervening claims'.[30] Many districts were remote and sparsely populated, and interstate and inter-urban rivalries held sway.[31] In South Africa, despite a strong federal lobby, white nationalism in 1910 had come, by way of the union of the four colonies, under a strong central government which, from the very outset, had an authoritarian cast.

Wildlife protection in Australia was the responsibility of each state, but each state also controlled its own land and resource utilization policy. This was not the case in South Africa, where wildlife protection was a provincial function, but land, agriculture, mining and other departments were in the hands of the central government. As well as demonstrating an outburst of national recon-ciliation, the foundation of the Kruger National Park in 1926 was precipitated by antagonisms between these various organs of central government.[32] Nothing comparable occurred in Australia; nor did wildlife protection give early direc-tion to the establishment of Australian national parks. Extinction of endemic species did not hold the same initial fear for Australians as it did for South Africans; as Dunlap argues, they considered their unusual creatures to be doomed to extinction anyway, because they could not compete against more advanced forms of life.[33]

Dongola Wild Life Sanctuary

While the foundation and management of the Kruger National Park was un-contested in white and state circles, the proposed establishment during the 1940s of a national park on the Limpopo River split the white population of South Africa. The reasons why Dongola became South Africa's first – and, to date, only – national park to be de-proclaimed are to be found in the fact that white landowning interests were threatened. Dongola would have been the first national park in southern Africa to have had landscape, archaeology and botany (not wildlife) as its core protectionist focus, and, in addition, it would have been the first trans-frontier park in the world. Called 'peace parks' today, these are now one of the major thrusts of the conservation endeavour in the sub-continent.

The national park which was proclaimed in 1947 and de-proclaimed a mere two years later highlights divisions within even the white 'nation' over national parks. Dr I. B. Pole Evans, a plant pathologist in the Department of Agriculture who took particular interest in indigenous vegetation, founded the Dongola Botanical Reserve in 1922. This consisted of nine adjoining farms, all 'worthless land' only 'fit for nature, nature left alone as man first saw it, in balanced equilibrium'. As no livestock was kept on Dongola, its pastures and springs remained abundant, soil erosion was minimal, and wildlife increased. In 1929 Pole Evans invited two National Party Cabinet Ministers (who were keen hunters) to his reserve, where they were impressed by the abundant wildlife and fine baobabs.[34] Their enthusiasm, together with that of Pole Evans's close friend Jan Smuts (who had founded the Rustenburg Game Reserve in 1909 and who was well known for his interest in nature and its conservation), spurred the botanist into a determination to transform Dongola into a national park.

Throughout the 1930s Pole Evans enlarged his botanical reserve to make it more worthy of national park status; his luck turned when his friend Smuts became Prime Minister in 1939, with Andrew Conroy as Minister of Lands. What Pole Evans had in mind, however, was an unusual national park for that time; he did not intend it to be a tourist centre, but an area dedicated to wilderness and scientific research. Moreover, it had an international dimension, because the neighbouring Rhodesian government and the chartered company of Bechuanaland agreed to co-operate in the venture.[35]

When the Dongola question was raised with the National Parks Board, its chairman, J. F. Ludorf, a National Party member, rejected the project out of hand.[36] Conroy went ahead nonetheless, announcing his intention in October 1944 to set aside 240,700 hectares as the Dongola Wild Life Sanctuary. Conroy's announcement unleashed a barrage of criticism. The local community in the northern Transvaal, principally Afrikaans-speaking, although led by the American-born manager of the Messina copper mine, conducted a concerted campaign against the proposed national park. Farmers in the area were joined by other government departments, organized agriculture, the press and the public. A Select Committee was appointed to investigate the matter and it sat for two years.[37] In Parliament, the 'Battle of Dongola'[38] generated some of the longest and most acrimonious debates up to that time. Dongola politicized nature protection to an extreme and bitter degree, and white opinion divided among party-political lines. An anti-imperial sentiment emerged, and the fact that Smuts was an 'internationalist', wanting to co-operate with Bechuanaland and Rhodesia, was held against him.[39] United States support for the project was also damaging, as was the expectation that it would be called the 'Smuts National Park'.[40] Another problem was that African opinion in the area was in favour of the Dongola scheme – anathema to the National Party.[41]

Despite these fierce objections, the Dongola Wild Life Sanctuary Act became law on 28 March 1947. But whereas the founding of the Kruger National Park had united people of different political persuasions, so divisive was Dongola and so intense the hatred it aroused, that it became an election issue in 1948. On coming to power thereafter, the new government immediately de-proclaimed the Dongola sanctuary.[42]

National parks: wildlife sanctuaries or game reserves?

It will have been noted that, although Kruger and Dongola were both established by the highest legislative authority in South Africa, one was called a 'national park', the other a 'wild life sanctuary'. Throughout the 1920s and 1930s, imperial protectionists struggled with the notion of a national park. The idea of permanence was attractive, but the word 'park' was problematic because it conveyed ideas of a public recreational playground. Also difficult, however, were the titles 'preserve' or 'reserve', with their connotations of exclusivity. James Stevenson-Hamilton, South Africa's leading wildlife conservationist,[43] complained frequently about the phrase 'national park'. In 1905 he called his game reserves 'game nurseries', as did Alfred Pease in 1908.[44] It seems that the contemporary terminology was fluid; in 1906 the Society for the Preservation of the Wild Fauna of the Empire called Yellowstone 'the great game reserve' of the United States.[45] Well into the late 1930s, Stevenson-Hamilton was calling for a 'National Faunal Sanctuary', a 'National Wild Life Sanctuary' or an 'animal sanctuary', attempting to promote an ideology of wilderness, 'roughing it' and allowing nature to take its course, rejecting 'park' both for its ideas of recreational comfort and scientific management intervention.[46] It was a losing battle, however; the word 'national', particularly after World War I, was enormously powerful in conjuring up regional pride.

As far as South Africa was concerned, its record from the 1920s to the 1940s was a mixed one. The Kruger National Park and Dongola Wild Life Sanctuary were divisive in different ways, but events in other protected areas, such as the abolition of a considerable number of provincial game reserves (together with immense wildlife extermination on the grounds of disease) and the establishment of very small species-specific national parks (Addo Elephant National Park, 1931, Bontebok National Park, 1931, Mountain Zebra National Park, 1937), occurred without overt national conflict. This was also when the huge Kalahari Gemsbok National Park was proclaimed (1931), with harsh consequences for the San people of the area. However, their resistance was localized and unpublicized and, by the 1960s, Dongola had been forgotten and Kruger and the other national parks were well enmeshed in apartheid South Africa.

The rash of national parks in South Africa in the 1930s may have been triggered by the surge of international protectionism at the beginning of the decade with the initiation of the American Committee for International Wild Life Protection and the growing imperial advocacy of either the establishment of a permanent international body for conservation or the use of the League of Nations for this purpose.[47] The London conference of 1933 was most successful, giving rise to a further conference in 1938.[48]

The Pilanesberg National Park

The 1960s saw the resurgence of international intervention, with the creation of a new, and different, national park in South Africa in response. While the Kruger National Park and the Dongola Wild Life Sanctuary consisted of reasonably unspoilt land and lay within the boundaries of the apartheid state, the Pilanesberg National Park owed its very origins to the 'homelands' policy of the Nationalist government of the 1960s. This national park, like Kruger before it, was a means of legitimizing a 'nation', this time of one of apartheid's 'independent states', and easing it into a place of international respectability.[49]

It would be inaccurate to think of the Pilanesberg National Park as a conserved natural area: it is more of a forced removal, land reclamation and game-stocking project. The idea of establishing some kind of wildlife tourist attraction in the unique Pilanesberg volcanic crater originated from the extremely conservative Potchefstroom University for Christian National Education in 1969, a time when self-governing and independent homelands were being mooted as apartheid blueprints for the future. Even in the early planning stage, opposition to the scheme from the Pilane clan was intense. Perhaps this local opposition had some effect, because it was not until the erection of the Sun City hotel, casino and entertainment complex nearby some years later that the national park idea became a reality.[50]

The traditional inhabitants of the Pilanesberg had been replaced by white owners fairly early in the colonial period, but legislation contained in the Native Trust and Land Acts of 1913 and 1936 meant that the area reverted to African ownership and white farmers were expropriated by the state. Little of the natural flora and fauna survived the wasteful pastoral practices of both whites and Africans. Roads, dams and other structures were sited unsuitably; invasive exotic plant species had taken hold; and soil erosion was considerable. However, the state was determined to forge a new land use for some 50,000 hectares of the crater and surrounding hills, and inhabitants were evicted in order to accomplish this goal. Little attention was paid to the concerns of the occupiers, the area merely being regarded by planners as 'a dusty valley of bankrupt farmers and jobless villagers'.[51] Vehement opposition efforts were unsuccessful and the entire Pilanesberg community was removed to Saulspoort.

Thus the Pilanesberg National Park had a difficult birth at a time when paramilitary wildlife management and anti-human ecology were powerful themes in national park dogma. Not only did local people oppose it, but the capital to begin land reclamation and game introduction projects was not forthcoming from the Bophuthatswana homeland. In the event, Anton Rupert's South African Nature Foundation[52] – whose close ties with the apartheid government are now well-established[53] – came to the rescue with R2 million, which 'drove out the livestock, pulled up the fencing, razed the farm houses and hovels and created a national park'.[54] Consultants were employed to plan the park and it was divided into sections for wilderness, hunting, and areas of differing tourist density. 'Operation Genesis' began and the translocation of wildlife commenced. Opposition from local people continued, however, and early management was corrupt and incompetent.[55] Moreover, scientists were concerned about the reintroduction of wildlife species from other areas of the sub-continent and the possible problems of mixing gene pools. When a newly introduced circus elephant killed a Brits farmer, scientists loudly denigrated the Pilanesberg Park.[56]

In time, however, experienced rangers, principally from Zimbabwe and Natal, were employed, land reclamation procedures began to take effect, and wildlife populations settled into their restricted habitat. Visitors – at first from Sun City, but subsequently from the Witwatersrand area as a whole – increased and the coffers of the park swelled. Announcements of the indirect benefits to be derived from tourism and increased wildlife husbandry became more strident and placated some of those whose removal had been necessary in order to nationalize the park land. Trespassing became less frequent and local neighbours began to benefit from culling programmes in the park. A portion of the fees charged for hunting – an unusual pursuit in a national park – was paid over to local councils. Environmental education programmes began and visitors were encouraged.

Conclusion: people and parks

After the Australian Nature Conservation Foundation was established, regular interstate meetings were held to co-ordinate national policy.[57] In 1975 the Australian National Parks and Wildlife Service was given jurisdiction over parks in the Northern Territory, a state in which Aboriginal land rights are a sensitive issue.[58] And thus, when the Pilanesberg was being constructed, the vast Kakadu National Park came into being in 1979 in an area which had been of special significance for Aborigines for 50,000 years and is rich in rock art and archaeological sites.[59] The lease arrangement between central government and Aborigines was a pioneering one, and although not without its management, legal and other problems,[60] co-operative ventures in Australia are providing examples for other countries, including South Africa, to follow.[61] A recent publication makes the observation that 'we are seeing in Australia a basic redefinition of what is a national park'.[62] Indeed, it is clear from the experience of Australia and elsewhere, that the hegemonic Yellowstone model of a wilderness free from people because it has been artificially 'created by bulldozers and fences, forced migration and resettlement' is inappropriate for the twenty-first century. Worse still, if adhered to, 'the consequences can be terrible'.[63]

In most parts of Africa, topics and policies of community involvement in protected areas are urgent issues at local, regional and state levels. National parks and other protected areas are now being used as tools for rural development and capacity building. While the national parks movement had federal ownership as its core concept, the notion of ownership is less central today. South Africa is attempting to create a fresh strategy,[64] although the legacy of apartheid weighs heavily and park managers are uncomfortable with challenges to their beliefs as to what constitutes a national park. It is not only former colonies with an underdeveloped population component that are participating in this reappraisal, however. So are the developed and densely populated countries, such as Britain, which have already addressed some of the difficulties of integrating people and nature.

'People and parks' is the new global national park priority; following the Bali declaration of 1982, protected areas must serve human society.[65] This focus is linked to a change in environmental ideology from preservation to sustainability on a planet of diminishing resources. It also has implications in that the protected areas are not only regarded as national institutions but as having international significance as well. Initiatives towards biosphere reserves and World Heritage sites should be seen in this context. The numerous international conferences on these subjects have resonances with the imperial conferences of the early part of the century. They bring with them, however, the possibility of a new kind of imperialism by way of intervention from a power base outside the region.[66]

Notes

1. Timothy Fritjof Flannery, *The Future Eaters* (Sydney: 1994), p. 5.
2. See, for example, W. C. Everhart, *The National Park Service*, 2nd edn (Boulder: 1983); Roderick F. Nash, *The American Environment: Readings in the History of Conservation*, 2nd edn (Reading, Mass.: 1976), 'The American Invention of National Parks', *American Quarterly*, 23/3 (1970), pp. 726–35, *Wilderness and the American Mind*, 3rd edn (New Haven: 1982); Alfred Runte, *National Parks: The American Experience* (Lincoln: 1979); John Ise, *Our National Park Policy: A Critical History* (Baltimore: 1961); William Beinart and Peter Coates, *Environment and History: The Taming of Nature in the USA and South Africa* (London: 1995).
3. J. G. Nelson, R. D. Needham and D. L. Mann (eds), *International Experience with National Parks and Related Reserves* (Waterloo, Ont.: 1978), p. 9.
4. Jean-Paul Harroy (comp.), *World National Parks: Progress and Opportunities* (Brussels: 1972), p. 97; Nelson *et al.*, *International Experience with National Parks*, p. 9.
5. Geoffrey Bolton, *Spoils and Spoilers*, 2nd edn (Sydney: 1992), pp. 2–21, 60; J. M. Powell, *Environmental Management in Australia 1788–1914* (Melbourne: 1976), p. 35.
6. Stuart A. Marks, *The Imperial Lion: Human Dimensions of Wildlife Management in Central Africa* (Boulder: 1984), p. 12.
7. Quoted in Luke Trainor, *British Imperialism and Australian Nationalism* (Cambridge: 1994), p. 73.
8. See Jane Carruthers, *The Kruger National Park: A Social and Political History* (Pietermaritzburg: 1995), and 'The Dongola Wild Life Sanctuary: "Psychological Blunder, Economic Folly and Political Monstrosity" or "More Valuable than Rubies and Gold"?', *Kleio*, 24 (1992), pp. 82–100.
9. David Western and Mary C. Pearl (eds), *Conservation for the Twenty-First Century* (New York: 1989), pp. 139–40.
10. John C. Freemuth, *Islands under Siege: National Parks and the Politics of External Threats* (Lawrence: 1991).
11. Particularly their control of the Australian landscape by fire: Bolton, *Spoils and Spoilers*, pp. 4–7; Stephen J. Pyne, *Burning Bush: A Fire History of Australia* (New York: 1991); Stephen Dovers (ed.), *Australian Environmental History: Essays and Cases* (Melbourne: 1994), p. 35; • Howard Morphy, 'Colonialism, History and the Construction of Place: The Politics of Landscape in Northern Australia', in Barbara Bender (ed.), *Landscape Politics and Perspectives* (Providence: 1993), p. 206.
12. TA CS396 10370/03, Transvaal Game Protection Association (TEPA) to Colonial Secretary, 18 November 1903; J. Stevenson-Hamilton, 'Game Preservation in the Transvaal', *Blackwood's Magazine*, March 1906, p. 409.
13. For details about this conference, see Public Record Office, London (PRO), FO403/302 7322; FO2/818; FO881 7394.
14. John M. MacKenzie, *The Empire of Nature: Hunting, Conservation and British Imperialism* (Manchester: 1988), and *Imperialism and the Natural World* (Manchester: 1990). See also William Beinart, 'Empire, Hunting and Ecological Change', *Past and Present*, 128 (1990), pp. 162–86.

15. Flannery, *Future Eaters*, p. 348.
16. William J. Lines, *Taming the Great South Land* (Sydney: 1991), p. 140.
17. Thomas R. Dunlap, 'Australian Nature, European Culture: Anglo Settlers in Australia', *Environmental History Review*, 17/1 (1993), pp. 27–8.
18. Tom Griffiths, *Hunters and Collectors: The Antiquarian Imagination in Australia* (Melbourne and Cambridge: 1996), p. 16.
19. J. M. Powell, *An Historical Geography of Modern Australia* (Cambridge: 1988), p. 40.
20. Tom Griffiths, '"The Natural History of Melbourne": The Culture of Nature Writing in Victoria, 1880–1945', *Australian Historical Studies*, 93 (1989), pp. 343–6.
21. Stevenson-Hamilton to Lagden, 4 September 1902 (TA SNA52 NA1904/02); Stevenson-Hamilton Documents in Trust, Reports by Ranger Gray, 1–7 October 1902 and 5–12 November 1902 (Kruger National Park Archives, Skukuza [KNP]); Report on the Sabi Game Reserve for the year ending August 1903 (TA SNA169 NA2063/03); Stevenson-Hamilton to McInerney, 18 August 1902; Native Commissioner Lydenburg to Secretary for Native Affairs, 18 November 1902 (TA SNA50 NA1751/02).
22. Stevenson-Hamilton to Secretary to the Administrator, 15 February 1911 (TA TPB784 TA3006).
23. *Annual Reports* of the Sabi and Singwitsi Game Reserves for 1912–16, 1918 (KNP).
24. Minutes of the National Parks Board, 4 April 1934, item 24 (KNP).
25. Minutes of the National Parks Board, 19 September 1949, item 3 (KNP).
26. *Annual Report* of the Sabi Game Reserve, 1905, item B57 (KNP); Acting Secretary for Native Affairs to Provincial Secretary, 1 April 1912 (TA TPS8 TA3072); *Annual Reports* of the Sabi and Singwitsi Game Reserves for 1913 (TA TPS8 TA3075); Acting Secretary for Native Affairs to Assistant Colonial Secretary, 16 May 1908 (TPS8 TA3072); *Annual Report* of the National Parks Board, 1932, p. 5 (KNP); Minutes of the National Parks Board, 19 March 1933, 3, 11 May 1938 (KNP); Warden to Secretary of National Parks Board, 30 October 1948 (K1/7-9-10, KNP); Central Archives, Pretoria (Cent.) NTS 2527 147/293 I, 1910–11; NTS 3589 853/308, 1930–9; NTS 7612 8/329, 1922; Minutes of the National Parks Board of Trustees, 1933 (KNP).
27. 'Verslag oor 'n ondersoek van die bestuur van die verskeie nasionale parke en die Nasionale Parkeraad se administrasie' (Hoek Report), September 1952, p. 136; Van Graan to the National Parks Board, 31 March 1950 (KNP K7/6 K7/7).
28. For example, R. Knobel, 'The Economic and Cultural Values of South African National Parks', in I. Player (ed.), *Voices of the Wilderness* (Johannesburg: 1979).
29. Jane Carruthers, 'Dissecting the Myth: Paul Kruger and the Kruger National Park', *Journal of Southern African Studies*, 20/2 (1994), pp. 263–83.
30. Powell, *Historical Geography*, p. 20.
31. Dunlap, 'Australian Nature, European Culture', p. 30.
32. See Jane Carruthers, 'Creating a National Park, 1910 to 1926', *Journal of Southern African Studies*, 15/2 (1989), pp. 188–216.
33. Dunlap, 'Australian Nature, European Culture', p. 32.
34. Dr I. B. Pole Evans (n.d., c.1944), 'The Dongola Wild Life Sanctuary Bill', especially sections 6 and 8 (Cent. LDE 1123 21439/29, vol. 2).
35. Cabinet memorandum, n.d.; Secretary for Lands to Secretary for Finance, 5 April 1942; Pole Evans to Secretary for Lands, 31 May 1943 (Cent. LDE 1123 21439/29, vol. 2).
36. Minutes of the National Parks Board of Trustees, 10 August, 13 November 1942; Report of the Select Committee, evidence of the Secretary of the National Parks Board, vol. 1, pp. 191–9 (KNP).
37. Union of South Africa, *Report of the Select Committee on the Dongola Wild Life Sanctuary Bill*, vol. 1, S.C. 12 (1945), vol. 2, S.C. 6 (1946).
38. *Pretoria News*, 8 September 1945.
39. Cabinet memorandum on proposed Dongola Nature Sanctuary (n.d.) (Cent. LDE 1123 21439/29, vol. 2).

40. House of Assembly, *Debates*, 4 April 1945, col. 4764; C. A. Heath to Secretary of the Interior, 11 September 1945, Secretary of the Interior to Heath, 27 September 1945 (National Archives of the USA [Washington], Record Group 79).

41. Native Commissioner to Secretary for Native Affairs, 2 July 1942 (Cent. LDE 1122 21349/29, vol. 1); Native Commissioner Louis Trichardt to Secretary for Native Affairs, 1 December 1944; memorandum 28 December 1944 (Cent. NTS 6998 291/321); House of Assembly, *Debates*, 5 April 1945, cols 4757–8; Dongola Wild Life Sanctuary Bill, 1945, Return of Replies (n.d.) (Cent. LDE 1123 21439/29, vol. 2).

42. For details following abolition, see Secretary for Lands to Secretary to the Treasury, 4 June, 28 September 1948, 23 August 1949 (Cent. TES 2628 F10/407); note, 28 May 1949 (LDE 1122 21439/13).

43. His publications include: *Animal Life in Africa* (London: 1912), *Wild Life in South Africa* (London: 1947), *The Kruger National Park* (Pretoria: 1928), 'The Management of a National Park in Africa', *Journal of the Society for the Preservation of the Fauna of the Empire*, 10 (1930), pp. 13–20, *South African Eden* (Cape Town: 1993); 'Wild Life Ecology in Africa', *Associated Scientific and Technical Societies of South Africa: Annual Proceedings, 1941–1942* (Johannesburg: 1942), pp. 95–106.

44. J. Stevenson-Hamilton, 'Game Preservation in the Transvaal', *Journal of the Society for the Preservation of the Wild Fauna of the Empire*, 2 (1905), pp. 20–45; Alfred Pease, 'Game and Game Reserves in the Transvaal', *Journal of the Society for the Preservation of the Wild Fauna of the Empire*, 4 (1908), pp. 29–34.

45. 'Minutes of Proceedings at a Deputation from the Society for the Preservation of the Wild Fauna of the Empire to the Rt. Hon. the Earl of Elgin, His Majesty's Secretary of State for the Colonies', *Journal of the Society for the Preservation of the Wild Fauna of the Empire*, 3 (1906), pp. 20–32.

46. Carruthers, *The Kruger National Park*, pp. 113–14.

47. H. R. Carey, 'Saving the Animal Life of Africa: A New Method and a Last Chance', *Journal of Mammalogy*, 7/2 (1926), pp. 79–81.

48. 'International Congress for the Protection of Nature', *Journal of the Society for the Preservation of the Fauna of the Empire*, 15 (1931), pp. 43–52; Agreement Respecting the Protection of the Fauna and Flora of Africa, Cmd 4453 (1933); International Convention for the Protection of the Fauna and Flora of Africa, Cmd 5280 (1936); American Committee for International Wild Life Protection, *The London Convention for the Protection of African Fauna and Flora* (Cambridge, Mass.: 1935).

49. *African Wildlife*, 37/6 (1983), p. 227.

50. M. R. Brett, *The Pilanesberg: Jewel of Bophuthatswana* (Sandton: 1989), pp. 111–12.

51. *The Star*, 11 December 1989.

52. The South African Nature Foundation has become the South African arm of the World Wildlife Fund.

53. Stephen Ellis, 'Of Elephants and Men: Politics and Nature Conservation in South Africa', *Journal of Southern African Studies*, 20/1 (1993), pp. 53–69.

54. *The Star*, 11 December 1989.

55. *African Wildlife*, 34/4 (1980), p. 18; *African Wildlife*, Jubilee Edition (1986), p. 13; Brett, *The Pilanesberg*, pp. 112–14.

56. Brett, *The Pilanesberg*, p. 112; *African Wildlife*, 37/6 (1983); *African Wildlife*, 43/6 (1989), p. 325. Attacks on tourists by relocated animals have continued in the Pilanesberg and elsewhere.

57. Harroy, *World National Parks*, pp. 100–2.

58. Dunlap, 'Australian Nature, European Culture', pp. 35–8; Lines, *Taming the Great South Land*, pp. 238–9.

59. Bain Attwood (ed.), *In the Age of Mabo: History, Aborigines and Australia* (St Leonards: 1996), pp. xxvii, 81, 147; Mary Blyth, John deKoning and Victor Cooper, 'Joint Management

of Kakadu National Park', in Jim Birckhead, Terry de Lacy and Laurajane Smith (eds),
● *Aboriginal Involvement in Parks and Protected Areas* (Canberra: 1992), p. 263.

60. Birckhead *et al.* (eds), *Aboriginal Involvement*; Sally M. Weaver, 'The Role of Aboriginals in the Management of Australia's Cobourg (Gurig) and Kakadu National Parks', in Patrick C. West and Steven R. Brechin (eds), *Resident Peoples and National Parks: Social Dilemmas and Strategies in International Conservation* (Tucson: 1991), pp. 311–33.

61. In South Africa these are called 'contract parks', of which the Richtersveld National Park was the first in 1991: see David Fig, 'Flowers in the Desert: Community Struggles in Namaqualand', in Jacklyn Cock and Eddie Koch (eds), *Going Green: People, Politics and the Environment in South Africa* (Cape Town: 1991), pp. 112–21.

62. Terry de Lacy, 'The Evolution of a Truly Australian National Park', in Birckhead *et al.* (eds), *Aboriginal Involvement*, p. 383.

63. David Foster, 'Applying the Yellowstone Model in America's Backyard: Alaska', in Birckhead *et al.* (eds), *Aboriginal Involvement*, pp. 363–4.

64. Robbie Robinson (ed.), *African Heritage 2000: The Future of Protected Areas in Africa* (Pretoria: 1995).

65. World Congress on National Parks, *National Parks, Conservation and Development: The Role of Protected Areas in Sustaining Society* (Bali: 1982); David Western and R. Michael Wright, *Natural Connections: Perspectives in Community-Based Conservation* (Washington, DC: 1994); N. Ishwaran, 'Biodiversity, Protected Areas and Sustainable Development', *Nature and Resources*, 28/1 (1992), pp. 18–25; West and Brechin (eds), *Resident Peoples and National Parks*.

66. For example, the CITES agreement: see Kevin A. Hill, 'Conflicts over Development and Environmental Values: The International Ivory Trade in Zimbabwe's Historical Context', *Environment and History*, 1/3 (1995), pp. 335–49.

CHAPTER 9

Scotland in South Africa: John Croumbie Brown and the roots of settler environmentalism

Richard Grove

The emergence of a critique of the environmental impact of settlement in the British colonial empire was pre-eminently a Scottish phenomenon, albeit influenced by some German and French lines of thought. The primacy of a Celtic ecological critique should come as no surprise. Despite King James and the accident of the Act of Union of 1707, Ireland, Wales and Scotland were England's first colonies and empire in a very real sense, subdued only by overwhelming military force, systematic land settlement, and the development of ideologies of occupation, survey and forced population removal. In considering the roots of settler environmentalism, the subjugation of the Scottish landscape and people under 'British' rule is especially significant. This is so not only because it brought wholesale changes in land ownership, major deforestation and population removal, but also because this brutal colonial imposition coincided temporarily with two powerful intellectual movements: the establishment of Scottish Enlightenment universities, where a brilliant flowering of intellectual life and training emerged; and the elaboration of an intellectually rigorous and socially vigorous Calvinist and Congregationalist Protestantism.[1] The mixture was an awkward but potent one. It gave rise, partly in response to the deforestation of Scotland by English soldiers and capital, to much of modern environmentalism as it has developed in the English-speaking world in the last two centuries, particularly in the English-speaking empires and neo-empires of Britain and the United States. By the mid-nineteenth century the mixture had been further enriched by the precocious development in Scotland, as in Holland and France, of romantic landscape tastes and by the development of a school of landscape painting. This paralleled the contemporary intellectual dominance of the Scots in all fields, especially economics, philosophy, history, botany, political science and engineering. But the lack of opportunities for employment for university-trained Scotsmen in their patrimony meant that their influence, especially in colonial employ, became disproportionately significant.

Some of the leading Scottish pioneer figures of the early environmental movement are now well-known, not least in the persons of John Muir in the

United States, and William Roxburgh, Alexander Gibson, Edward Balfour and
Hugh Cleghorn in India. Before 1900, possibly only George Perkins Marsh in
the USA and Baron Ferdinand von Mueller in Australia approached these
Scotsmen in power and actual influence in the Victorian environmental movement.[2] My focus here is on the life and work of the Reverend John Croumbie
Brown (1808–94).[3] Brown was employed by the Cape Colony government
from 1862 to 1866 specifically to examine and advise on the biological aspects
of colonial settlement; he was a highly critical and rigorous commentator on
the processes of ecological change. He became the single most influential voice
in the formation of a colonial and North American discourse on forestry,
irrigation, range management and on the environmental impact of settlement.
The strength of his critique depended on his very wide scale of reference both
to published authorities and to detailed case-studies from all parts of the world
and from many historical periods, as well as on his grasp of global processes.
His critique resulted in voluminous publications.[4] Much of the interest in
examining the life of Brown, however, lies in his antecedents and in an explanation of why a missionary from a stern and staid Calvinist background should
have become such a radical and innovative critic of the ecological activities of
the colonial state and even have been sacked for his extremist views. An
interpretation of the intellectual forces at work in Brown's life, as it expressed
itself in his career in the Cape Colony and in his later publications, stands as
a very useful guide to the roots and strength of Scottish environmentalism in
general, as it became the master discourse of colonial and post-colonial environmental sensibilities.

   In 1894 Henry Rider Haggard published one of his lesser-known novels. It
was called *The People of the Mist* and was strangely reminiscent of *The Lost
World*, the influential fantasy written by that great Scottish doctor and enthusi-
ast of Indian tribal culture, Sir Arthur Conan Doyle. Like Conan Doyle, who
had never been to Roraima in Venezuela, Haggard had never visited the Mulange
Mountains which had inspired the *The People of the Mist*. Indeed, no European
climbed Mulange until Alexander Whyte did so in 1896 and discovered the
giant cedar, *Widdringtonia whytei*, the great redwood of southern Africa.[5] But
Haggard's *The People of the Mist* was remarkable for combining two entirely
different sets of myths about people and landscape. He had heard Lomwe and
Yao stories about Bushmen who dwelt in the recesses of Mulange itself and
possessed a powerful and deathly magic against outsiders.[6] But he had also read
Sir Walter Scott's *The Legend of Montrose*, in which Scott's so-called 'Children
of the Mist' were the surviving remnants of bands of desperate highland natives,
willing to die rather than face capture at the hands of the lowlanders. One
might think such a confusing mix of Scottish and African legend would be
merely Haggardian eccentricity. But, on the contrary, Haggard trod in the steps
of a Scottish tradition of identifying with the indigenous African, oppressed in
their highland fastnesses, that was nearly a century old.[7] It was also a tradition
vital to understanding the whole psychology of Scottish settler culture and

environmental attitudes in southern Africa. Edward Said and Mary Louise Pratt have bidden us see the world in terms of settler 'us' and colonized 'other'. As Nigel Leask and John MacKenzie have recently reminded us, however, the reality is – at the very least – one of 'this', 'that' and 'the other'.[8] And it is the nature of the Scottish 'that' which I will discuss here.

Haggard's 'mountains of the mist' are a case in point. By 1896 they stood in the Central African Protectorate, later the Nyasaland Protectorate. This was in some senses a Scottish colony, whose birth had been actively delayed by an English Prime Minister.[9] But was it a settler colony or not? It is not at all clear that one can meaningfully separate settler colonies from other kinds of colonies. In terms of the evolution of the landscape, British colonies were all settler colonies. Moreover, it is difficult to separate the impact of western economic penetration from a specifically colonial economic impact in a settler colony.

In *Green Imperialism* I tried to indicate how the colonial context stimulated the development of an environmentalist critique and conservationist ideology in what were, in the first place, small island settler colonies.[10] A distinctively French ideology of conservation drew heavily on English desiccation ideas and was strongly tied to social reformism in proto-revolutionary and anti-urban movements (especially on Mauritius), and to a cult of the South Sea island. The forest conservation movements in India and the United States were, I argued, essentially derivative, even though the ideas of their Scottish proponents were quite clearly distinctive, romanticist and anti-establishment. In South Africa, however, the Cape always posed more of a problem, not least because it was a context so clearly dominated by Scottish thinking. My earlier work has explored conservationism in South Africa as primarily a product of evangelical notions.[11] But this is far too limited an explanation, because it fails to take into account the social context of Scottish emigrants. Something far more important was actually taking place both in the evolution of environmentalism and in the evolution of the Scottish nation itself; both developments crystallized in the context of colonial settlement in southern Africa and, to a much lesser extent, in Canada, the West Indies, India and Australia.

What emerges in the environmental history of the Cape is a quite specific linkage between a critique of the social and ecological impact of settler rule and a nascent Scottish environmentalism which was not only largely indigenous to Scotland in its roots, but was itself nurtured by a social response to post-1707 colonial rule in Scotland. There, evolving landscape and environmental sensibilities were becoming a major vehicle for the expression of a national identity constructed and asserted against Englishness and English rule. This aesthetic sensibility was paralleled by a mythologization of Scottish history, folklore, and language which built upon the already sharply distinctive religious tradition of the new colony of Scotland in an attempt to regain the lost world of pre-Union times. The most important regional roots of this striving for Scottish identity can be pinpointed in the Scottish border country of East Lothian (or Haddingtonshire) and in the counties running south to the border, in Teviotdale and in the 'Waverley' country around Kelso. The intellectual superiority of the Scottish universities and their place in the mainstream of the European Enlightenment were already a sharply distinguishing factor in the emergence of a new identity. But Scotland was also looking in its own very specific geographical directions. The pressure necessarily placed on Scottish graduates, especially doctors and clerics, to seek employment outside Scotland had already helped to develop a national obsession with Africa, to the point that the continent had become a part of popular culture by 1800. Popular acclaim for the life, travels and tragic end of Mungo Park (1771–1806) on the Niger meant that he acquired a sainted fame and many enthusiastic disciples ready to emulate his life – and perhaps even his martyrly death – in southern Africa.

By the 1840s, popular literature had massively reinforced the romanticization of Africa in the Scottish mind and had provided a context in which Scottish ambitions could outshine the English. The *Penny Magazine*, a journal with a very high circulation in the first decades of the nineteenth century, set the tone of this romanticization. Among its most loyal contributors in the 1830s was Thomas Pringle.[12] The sheer bulk of articles by him and others on animals, landscape and people in southern Africa ensured that young aspiring Scotsmen would, when thinking beyond Britain, think first of southern Africa as a new frontier for adventure, money and mission. Moffat, Livingstone and John Croumbie Brown were amongst them. But Thomas Pringle, who arrived in the Cape in 1819, was undoubtedly a most perceptive observer when it came to articulating the critical difference between the Scottish and the English response to the Cape landscape. As his ship stood off the Cape at Simonstown, Pringle recorded:

> The sublimely stern aspect of the country so different from the rich tameness of ordinary English scenery, seemed to strike many of the Southron [English] with a degree of care approaching to consternation. The Scotch, on the contrary, as the stirring recollections of their native land were vividly called up by the rugged peaks and shaggy declivities of this wild coast, were strongly affected, like all true mountaineers on such occasions. Some were excited to extravagant spirits, others silently shed tears.[13]

Pringle recognized and represented a culture, in his *African Sketches*, that differed from the English in its readiness to incorporate and honour the verna- cular and the indigenous. Scottish culture was already heterogeneous and, therefore, had no difficulty in continuing to be so in the Cape. Pringle explicitly identified the oppression and slavery of the African indigene with the oppression of the Scots. He cultivated a constant pairing not only of Scots with African cultures, but of Scots with African landscapes. The two were completely inter- twined. Since landscape had already been sacralized as a leitmotif of Scots identity, it was but a short step to romanticize and even sacralize the Cape landscape, which Pringle and his imitators, including John Croumbie Brown, very soon did. The process was made easier by the way in which the Scots settlers of 1819 baptized their new hills with the ancient names of the borders: Eildon, Teviot, Tyne, etc. In his poems, Pringle conjured up intensely visual images linking and conflating the longed-for landscapes of the Cape and the Scottish borders until they were semantically indistinguishable. Their misty recesses sheltered tribes of indigenes, Hottentot or highlander, all equally harried by the English invaders.

Pringle became acutely sensitive both to the ill-treatment of the Africans by the English and to the growing tendency of the Cape government (under Governor Somerset) to censor his own outspoken journalism. This meant that he gradually evolved, in his poems and popular writings, a concept of the Cape as a moral landscape crying out for reform or development, in which the harm done to Africans had to be expiated by the European, to the point even of arming them for rebellion. Indeed, this was a view widely held by evangelicals in the 1830s. The concept of a moral landscape or moral wilderness soon attached itself very directly to the state of land, forests and landscape and became a ruling idiom in the opinions and writings of Pringle, Livingstone and their many Scottish emulators. The moral purposes of the Lord would thus be worked out physically, in the landscape itself.

The energy with which John Croumbie Brown sought to redeem the 'moral wilderness' of the Cape, as he called Cape society in his sermons in the mid-1840s, reflected the terminology and idioms of Pringle.[14] Deforestation and veld-burning were evils which could be cast out only through tree-planting and irrigation. But it was a redemptive pastoral discourse that acquired sufficient momentum to affect colonial land and forest policy from 1872 (when Joseph D. Hooker, the Director of the Royal Botanical Gardens at Kew, took up Brown's ideas) until the twentieth century, and by no means only in the so-called settler colonies.[15] Moreover, Brown shared Pringle's worries about the threat to indigenous peoples, although he did not always express them explicitly. In the introduction to his first conservationist text, *The Hydrology of South Africa*, Brown referred pointedly to the work of Fenimore Cooper, author of *The Last of the Mohicans*, possibly making a subliminal connection between a threatened American indigenous culture and extinctions threatening other cultures, even human culture itself. This was a notion that he shared with Elias de Fries, the pioneer Swedish ecologist from whom Brown derived so many of his ideas about the impact of modern man. For Brown, however, the survival of trees or forests themselves, not just the landscape, served to totemize survival or moral quality. This was integral to the preservation of a 'territory', a religious 'promised land' which he referred to specifically as 'Judah' or, sometimes, 'Judaea'.

Simon Schama has recently demonstrated the connections between the desire for the survival of large native trees and the forging of an American nationalism and identity in the West in the 1860s. Similar meanings attach to large trees in Australia.[16] As John Croumbie Brown recorded in 1881, Walter Scott made very specific claims for the sylvan character of Scotland in a pre-British past:

'History, tradition and the remains of huge old trees and struggling thickets, as well as the subterranean wood found in bogs and mosses, attest the same indubitable fact.'[17] But Scottish claims and connections to landscape in the context of settlement, while analogous to those of American or Australian nationalism, cannot be disentangled easily from specifically religious notions of redeeming or replanting the Judaea of the colonized 'promised land' and saving its people – indigenous or settler – from damnation or extinction. The ambitious task of preventing desiccation was a major motivator for arboreal redemption and reafforestation. But the link between a Cape threatened with environmental ruin by a profligate settler population and a Scotland already made barren by the evils of the English was always present. For Brown, there were sound historical reasons for making these connections, including precedents from his own home region of Haddington for redeeming the damage.

There seems little doubt that the accelerating rate of deforestation in much of Scotland after the Union, aided by deliberate military burning and the advent of the iron-smelting industry, made a great impression on contemporaries. A number of useful insights in this respect are left for us by Boswell and Johnson in their narratives on a tour to the Western Isles in 1773. Johnson, in particular, was horrified by the barrenness of the landscape and the paucity of tree cover, associating the very barrenness with the intrinsic inferiority and primitiveness, as he saw it, of Scottish society. By contrast, the Scottish Boswell and his contemporaries, particularly Lord Monboddo (the 'Scottish Rousseau', as he was often called) actually confessed to a yearning for primitive and untouched peoples and landscapes, especially those of the islands of the South Pacific. Boswell, who knew Captain Cook, expressed a wish on one occasion to go and live for three years in Otaheite, in order to meet with people unlike any that had yet been known 'and be satisfied what nature can do for man'.[18]

Rousseau was not widely read or known in England at this time. Quite the contrary was the case in Scotland, however, and foremost among his students was James Burnett, Lord Monboddo.[19] Monboddo was a pioneering anti-modernist who actively despised all forms of war, modern transport, food preparation and estate management. He also espoused both nudism and vegetarianism. He chose to allow his estates to grow naturally as they would, sensitive to the devastation that had been wrought by Monk's armies and those of many other European campaigns. Monboddo's nihilistic and somewhat pessimistic naturalism stands in contrast to the emerging ideas of many of his contemporary landowners and of many Scottish botanists of the period, who increasingly saw tree-planting as a useful and redemptive activity, embarrassed and spurred on as they were by the derisive remarks of English commentators such as Johnson. From at least as early as the end of the second Jacobite rebellion, many Scottish landowners, under a *pax Anglica* – and particularly those in Fife, Aberdeenshire, Perthshire, East Lothian (Haddingtonshire) – embarked on campaigns to develop tree plantations on their estates. Among the principal advocates of this new enthusiasm to reclothe the Scottish countryside after the ravages of the subjugation of the Jacobite rebellions were four successive Earls of Haddington whose lives spanned the late eighteenth, nineteenth and twentieth centuries: Thomas Haddington, the eighth earl; Thomas Haddington, the ninth earl; his nephew, the tenth earl, and *his* son in turn, George, eleventh earl of Haddington. From the first Thomas onwards all were well-known agrarian improvers and foresters. They carried out extensive tree-planting around the town of Haddington and wrote several treatises upon the subject, always utilizing the latest French literature and, in turn, influencing later French tree-planting practice.[20] Indeed, the silviculture of the Earls of Haddington was the main inspiration for the extensive *reboisement* of south-western France, especially of the shifting sand-dunes of Les Landes, and, subsequently, of many British and French colonies.[21]

It was at Haddington that John Croumbie Brown was born in 1808, in a setting at the centre of the Scottish tree-planting movement, only a few miles from Dunbar, the birthplace of John Muir.[22] Like Muir, Brown was the offspring of an intensely religious and Calvinistic family; he was the grandson of the Reverend John Brown (1722–87), the renowned author of the 'Self-interpreting' Bible, which went into fifty-one editions. This family tradition of frequent publication and revision and of preaching and interpreting a religious tradition according to one's own lights and observations, was to be vitally formative in Brown's own work. The inspiration for the preaching of the Word was to be found and adapted in the immediate life and circumstances of the preacher and his congregration.

Brown had originally intended, like Livingstone a little later, to devote his missionary inclinations to Africa. Instead, however, the London Missionary Society sent him in 1833 to nurture a small congregation in St Petersburg in Russia, a post which he retained for four years. In 1844 he was finally sent to the posting of his choice in the Cape, with the aim of making another Congregationalist community self-supporting. We may suppose that Brown had been impressed by the huge and largely undisturbed forests in the regions adjacent to St Petersburg, and the shock of arriving in a largely deforested Cape must have been a sharp one.[23] Whatever the roots of his inpiration, we know that by early 1845 Brown was preaching sermons in his church in Cape Town with a quite explicitly environmentalist message.[24] Some of these were published as a book in 1847 called *Pastoral Discourses*. As his text for Sermon 8 in 1846, Brown chose Isaiah 32: 13–15:

> Upon the land of my people shall come up thorns *and* briars; yea, upon all the houses of joy *in* the joyous city: Because the palaces shall be forsaken; the multitude of the city shall be left; the forts and towers shall be dens for ever, a joy of wild asses, a pasture of flocks; Until the spirit be poured upon us from on high, and the wilderness be a fruitful field, and the fruitful field be counted for a forest.[25]

Brown stated that his preferred exposition of this prophecy was that referring to the invasion of Judaea by Senanacherib, the severe sufferings of the people during invasion, their deliverance by the Lord and their subequent prosperity. Brown continued:

> if then I were asked what I consider the means most likely to bring under cultivation the vast uncultivated deserts of this land the waste-howling wilderness stretching away from this point to the far north I would reply at once:– The spread of pure religion; then shall the earth yield her increase ... so long as one sows and another reaps, few will be willing to till the ground.[26]

Later, Brown asks the question:

> [if I were asked to] consider the means by which the moral wilderness be reclaimed I would reply at once ... by the spirit of God being manifested by his people ... Let them act as stewards of the manifold Grace of God ... while they instruct the ignorant ... and that which is now a moral waste will be seen to rejoice and to blossom as the rose, being covered with trees of righteousness, the planting of the Lord, and He will be glorified ... they shall revive as the corn and grow as the vine; their branches shall spread and their beauty shall be as the beauty of the olive tree, and the scent thereof shall be as the Vine of Lebanon.[27]

Brown chose to prescribe in verse the 'work' that needed to be done to achieve this:

> Through the desert God is going
> Through the desert waste and wild:
> Where no goodly plant is growing,
> Where no verdure ever smiled;
> but the desert shall be glad;
> And with verdure soon be clad.
>
> Where the thorn and briar flourished
> Trees shall be seen to grow
> Planted by the Lord and nourished
> Stately, fair and fruitful too:
> See! they rise on every side:
> See! they spread their branches wide.
>
> From the hills and lofty mountains
> Rivers shall be seen to flow,
> There the Lord will open fountains;
> thence supply the plains below.
> As he passes, every land
> Shall confess his powerful hand.[28]

In 1847, the year in which *Pastoral Discourses* was published, Brown toured the Cape on horseback, travelling across the Karoo during drought. The severity of the conditions and the sufferings caused by water shortage so impressed him that he was struck by the need to convert the vivid language of his ministry into a practical programme. 'On my return to Cape Town', he wrote, 'I communicated to others the impressions I had received of greatly modifying the effects there produced by the aridity of the climate.' [29]

In 1848 Brown returned to Scotland. Frequent changes of scene allowed Brown both to compare landscapes geographically and to observe rates of landscape change, particularly at the Cape, to which he was to go back in 1862. On his return to Scotland, Brown chose, significantly, to move to Aberdeen. In this ancient and intellectually stimulating locale, the Alma Mater of many Scottish colonial doctors and botanists, Brown was able to pursue his neglected scientific interests and to teach at King's College, Aberdeen, where he was appointed lecturer in botany in April 1853.[30] This appointment allowed him to pursue his now very practical interest in arboriculture.

In 1862 Brown moved back to the Cape Colony as Colonial Botanist, the successor to Dr Ludwig Pappe, who had been appointed as the first Colonial Botanist in 1858. Brown's appointment seems to have been made at the behest of Saul Solomon, the coloured Cape Town publisher who had published *Pastoral Discourses*. Brown arrived in the colony in the worst of a run of drought years from 1861 to 1863, which we now know were closely associated with the very severe ENSO event which peaked in 1862.[31] The drought clearly had a colossal impact on his already alert perceptions and made him fearful of the supposed process of 'desiccation' (as he termed it).

Brown's responses to the developing ecological crisis of the Cape while he was Colonial Botanist can be charted through his official reports, letters and later books and articles. Prior to 1864, it appears that Brown felt very much at sea with regard to serious authorities on the subject of ecological change. But in the decade preceding the publication of his *Hydrology of South Africa* in 1875, Brown organized his ideas and quickly grew more confident in attributing the physical and moral causes of environmental decline in the Cape. Perhaps surprisingly, he accepted Darwin's new theories of natural selection, using them to argue for the introduction of exotic and 'more successful' drought-resistant tree species into the Cape. But he was given far more ammunition by G. P. Marsh's *Man and Nature* (1864), which contained solemn and doom-laden warnings, based on historical lessons from the Levant and Ancient Greece and Rome, about the climatic effects of deforestation. Marsh was also familiar with the work of the early Indian environmentalists, particularly that of Hugh Cleghorn, who had published his *Forests and Gardens of South India* in 1861. This last work, hostile to both industrial deforestation and to shifting cultivation, clearly impressed Brown.

Alongside the influence of these new authorities, however, we also need to understand why Brown wrote as many books as he did and what he saw as his purpose in doing so. He believed, like his grandfather, in spreading the Word through print, or through any medium of communication, preferably as effective and modern as possible, in order to perform a missionary task, not

necessarily a solely religious task. Brown had defined his terms in his first
non-religious book, a translation of the works of the Reverends T. Arbousset
and F. Daumas of the Paris Missionary Society, two missionaries who, like
Livingstone, combined gospel with exploration.[32] The young Brown had met
Arbousset in Cape Town in 1846 and it seems that he may have decided to
model himself upon his French colleague as well as to translate the *Narrative*
of Arbousset's travels (which had commenced in 1836). In 1851 Brown ration-
alized the publication of the translation on the grounds that:

> it embodies information of considerable importance to all who are interested
> in the Settlement of Natal; – but chiefly because circumstances have arisen
> which render it desirable that correct information with regard to tribes visited
> by him should be in possession of those who are interested in the Aborigines
> of the country.

Brown went on to remind his readers, in what was the year of the Great Exhibition in London, of other missionaries who had performed well in the scientific
field. The most important achievements of his French colleagues, Brown believed,
were to uncover new ethnological data on previously 'unknown' and possibly
threatened highland peoples and to discover the sources of the principal rivers of
South Africa. The preoccupation with water sources flagged what was already
an obsession with Brown, the provision of water in a progressively parched land,
as well as its corollary, a concern about the effects of incendiarism on vegetation and water supply. Spreading the Word and spreading the water of life in an
arid land were naturally connected notions for Brown.

Between 1862 and 1866, Brown had tried to spread his concerns about
deforestation, incendiarism and irrigation – some of which originated in the
work of his predecessor, Ludwig Pappe, and some which were his own –
through the mundane means of the colonial blue books and government reports
of the Cape Colony government, and through committees of opinion-formers
and legislators in the Cape Parliament. He supplemented this 'missionary' work
with a voluminous correspondence with members of the colonial settler community of all walks of life, with a catholicity and indiscriminate social eye that
shocked many of his staid contemporaries in Cape Town. Some of this correspondence was printed alongside his official reports as Colonial Botanist during
the 1860s, but most of it became raw material for later conservationist and
forestry tracts. In his official capacity, Brown travelled extensively to lecture
to farmers and settlers on his pet subjects of deforestation, irrigation and soil
erosion. But his critical approach to European agriculture and industrial forest
clearance was already raising worried eyebrows – of the kind that led, ultimately,
to his sacking. Brown's main concern was always to take his heartfelt message
ever farther and deeper and to reach more ears. In fact, his colonial reports had
already alarmed and stirred into action some very influential figures, not least
Sir Joseph Hooker, Director of Kew Gardens, who, from 1872, used Brown's
official reports on the degradation of the Cape environment as a scientific stick

with which to beat colonial governments on the subject of forest policy through-out the world. Much of the interest of metropolitan botanists in the state of the Cape environment derived from the diversity and splendour of its flora and its *fynbos* plant kingdom, a uniqueness already recognized by Sir William Hooker, father of Joseph and also a Director of Kew, who had taken a particular interest in forest preservation at the Cape. This external support, while tempor-arily helpful to Brown, counted for surprisingly little at the local political level in the Cape.

Brown was peremptorily dismissed from the employ of the government of the Cape Colony in 1866, largely because of his public condemnation of the illicit deforestation carried out by railway interests with influential friends in Parliament. He returned to Scotland, where he continued to seek ways of diffusing his increasingly well-informed views on environmental degradation and the need for state intervention, a conviction which expressed itself in a focus on the need for state forestry training.[33] Brown was much struck by the idea of a 'postal' or 'open' university that was, at the time, being proposed for the enlightenment of far-flung Australian settlers. He saw this as a way of communicating his views on state intervention in matters of environmental degradation. The proposal had first become known to him through an article in the *Spectator* about a suggestion made by George Baden-Powell, brother of an Inspector-General of Forests in India. In India, agricultural extension under a Board of Agriculture had started in 1879, after the appalling ENSO-caused droughts of 1877–9. In both Australia and South Africa, the problem of pros-ecuting meaningful agricultural improvement of the kind increasingly being advocated by state-employed 'experts', especially with regard to irrigation, range management and veterinary science, needed to be considered in the context of very large semi-arid regions where communications were extremely difficult.[34] Baden-Powell praised the new University of Sydney, with its nice new buildings and its position on a high eminence overlooking the city. But, Baden-Powell continued, it is 'impossible to get a real University education' at Sydney, since 'New South Wales, though fortunate in possessing the services of several clever professors, is as yet too young to provide a sufficiently large number of undergraduates.' In short, Baden-Powell said, a university was not relevant to the severe problems of settler agriculture, with its highly dispersed but youthful population. Instead, he noted:

> the happy idea of starting what may be called a Postal University has lately claimed attention. Such a method of education should prove an immense boon to Australia. Thousands and thousands of miles of bush country are held by squatters, and their work, necessarily of a very isolated character, is mainly in the hands of young men, most of whom have entered on its life fresh from school and full of aspirations. At times they have a great amount of leisure on their hands and this a very large majority of them would be very glad to devote to self-improvement.[35]

Baden-Powell's idea was clearly extremely appealing to anyone who saw the business of responsible resource management by settlers as being an article of faith, or, indeed, a message to be preached to a whole people, elect or not!

After seeing this article, Brown had written a letter to the editor of *The Colonies*, proposing that the journal might like to partake in a postal university scheme in which his own treatises on various subjects might be specially printed and transmitted on light paper to isolated farmers in several colonies. Brown had Australia and South Africa specifically in mind. Sadly, *The Colonies* chose not to take up Brown's adoption of Baden-Powell's revolutionary open university of natural resource management. The concept did, however, encourage Brown to embark on a grand scheme of natural resource treatise-writing for educational purposes, to which he adhered for the next twenty years. Brown's original schema for a global series of books on forest and natural resource problems, first published with *The Hydrology of South Africa* in 1875, makes fascinating reading, not least in indicating which projects he eventually decided *not* to tackle.[36]

The Hydrology of South Africa represented Brown's first shot at what became a distinctive genre. It was 'designed primarily for the benefit of the colonists at the Cape of Good Hope, but it embodies information which may be useful in other lands beside South Africa'. Brown deliberately quoted at great length from his widely culled and eclectic sources, for the sake of colonists who could not reach libraries. Brown tells us: 'I have availed myself gladly of permission given to me by the Hon. George P. Marsh, minister of the United States of America at Rome, to quote freely from his valuable work, *The earth as modified by human action.*' But he also mentions what was ultimately to be a much more important and pre-Marshian source, the long paper written by J. Fox Wilson for a meeting of the Geographical Society in 1864 on 'the desiccation of the valley of the Orange river'.[37] 'I have also quoted largely from the works of Dr Livingstone and others', Brown tells his readers.

Both Wilson and Livingstone posited the long-term desiccation of the whole of southern Africa (and, implicitly, other equally arid areas of the world), although they used different mechanisms to explain the phenomenon as they saw it. Brown was thoroughly convinced by the phenomenon of desiccation, having observed, as had both Wilson and Livingstone, the extraordinary severity of the 1862 El Niño drought and the mortality that accompanied it. The apocalyptical results of this drought, as well as the terror of the seasonal floods that followed it, run like a wakeful ghost through the rest of Brown's writing career. Indeed, the drought – and the progressive and artificially induced desiccation process which Brown believed it evidenced – provided a 'scientific' backbone to a semi-religious discourse. In this discourse, the sins of the profligate colonial settler threatened the whole of a broader society (settled in a new 'Judaea') on behalf of which Brown had set himself up as environmental reformer and struggler. Accordingly, irrigation, forest protection, fire avoidance and reafforestation emerged as the new 'good works' of the colonial pilgrim.

The first two books in Brown's grand schema (one is tempted to say 'the first two books of Brown') constitute a deliberately constructed and Dante-like polarity. The first paints a picture of possible damnation through human-induced desiccation, in which the erosive sins of the colonial settler are itemized in a historical format, while the second book offers the possibility of redemption through *reboisement*, almost as an arboreal Paradiso offered to redeem a desiccatory Purgatorio. Moreover, Brown actually advertised *Reboisement* in the back of the first edition of *The Hydrology of South Africa* with a four-page sermon urging the colonist both to buy the book and to start planting trees before an environmental perdition should overwhelm him.

Brown devoted the rest of his life to writing out the tablets of a covenant of environmentalism deeply inspired by a resurgent Scottish and religious identity, urged on by his observations of the reality of the fragile Cape. He acquired many disciples, too, including Franklin Benjamin Hough, the founder of the American Forest Service, who stayed with him for several months in 1883 and toured Scotland's forests with him. Hugh Cleghorn, the second Inspector-General of Indian Forests, also became an admirer and shared many speaking-platforms with him. Ultimately, Brown's writings focused on a series of books calling for the establishment of schools of forestry to serve Britain and the empire. This was a perfectly rational and utilitarian message. But another reading suggests that Brown was trying to hand on to posterity the message of his *Pastoral Discourses* of forty years earlier. He was advocating institutions from which his prescriptions for planting the 'trees of righteousness' might continue to thunder down the years and across the semi-arid landscapes and moral wildernesses of the tropics, just as his words had thundered from a Cape Town pulpit with all the energy of Scottish ecological redemption.

Notes

1. A. Chitnis, *The Scottish Enlightenment* (London: 1976).
2. Richard H. Grove, *Green Imperialism* (Cambridge: 1995).
3. For an earlier treatment of the history of Cape environmentalism, see Richard Grove, 'Early Themes in African Conservation: The Cape in the Nineteenth Century', in David Anderson and Richad Grove (eds), *Conservation in Africa* (Cambridge: 1987) pp. 21–39.
4. John Croumbie Brown's writings include: *The Truth and Truths of Christianity* (Cape Town: 1845), 153-pp.; J. T. Arbousset and P. Daumas, *Narrative of an Exploratory Tour to the North-East of the Colony of the Cape of Good Hope*, ed. and trans. John Croumbie Brown (1846; Aberdeen: 1852); *Pastoral Discourses* (Cape Town: 1847), 256-pp.; *Report of the Colonial Botanist* (Cape Town: 1863); *Report of the Colonial Botanist* (Cape Town: 1865), 120-pp.; *Report of the Colonial Botanist* (Cape Town: 1866), 15-pp.; *Circular from the Colonial Botanist Relative to South African Plants Desired by the Directors of Botanic Gardens in Europe and Elsewhere* (Wynberg: 1866), 6-pp.; *The Hydrology of South Africa; or, Details of the Former Hydrographic Condition of the Cape of Good Hope, and of Causes of its Present Aridity, etc.* (Edinburgh: 1875), 260-pp.; *Reboisement in France: or, Records of the Replanting of the Alps, the Cevennes and the Pyrenees with Trees* (London: 1876), 351-pp.; *Forests and Moisture: or, Effects of Forests on Humidity of Climate etc.* (Edinburgh: 1877), 308-pp.; *The Schools of Forestry in Europe: A Plea for the Creation of a School of Forestry* (Edinburgh:

1877), 72-pp.; *Water Supply of South Africa and Facilities for the Storage of it* (Edinburgh: 1877), 651-pp.; *Pine Plantations on the Sand Wastes of France* (Edinburgh: 1878), 172-pp.; *The Forests of England, and the Management of them in Bye-gone Times* (Edinburgh: 1883), 263-pp.; *French Forest Ordinance of 1669: With Historical Sketch of Previous Treatment of Forests in France* (Edinburgh: 1883), 180-pp.; *Finland: Its Forests and Forest Management* (Edinburgh: 1883), 290-pp.; *Forests and Forestry of Northern Russia and Lands Beyond* (Edinburgh: 1884), 279-pp.; *Forestry in Norway, with Notices of the Physical Geography* (Edinburgh: 1884), 227-pp.; *Forestry in the Mining Districts of the Ural Mountains in Eastern Russia* (Edinburgh: 1884), 182-pp.; *Introduction to the Study of Modern Forest Economy* (Edinburgh: 1884), 228-pp.; *Forests and Forestry in Poland, Lithuania, the Ukraine, and the Baltic Provinces of Russia* (Edinburgh: 1885), 276-pp.; *School of Forest Engineers in Spain Indicative of a Type for a British National School of Forestry* (Edinburgh: 1887), 232-pp.; *Schools of Forestry in Germany, with Addenda Relative to a Desiderated British National School of Forestry* (Edinburgh: 1887), 232-pp.; *Management of Crown Forests at the Cape of Good Hope, under the Old Regime and the New* (Edinburgh: 1887), 352-pp.; *Centenary Memorial of the Rev. John Brown, Haddington; a Family Record, Compiled by his Grandson, J. C. Brown* (Edinburgh: 1887), 223-pp.; *People of Finland in Archaic Times: Being Sketches of them Given in 'Kalevala' etc.* (London: 1892), 290-pp.

5. The Mulange Mountains lie in the south-east corner of today's Malawi, on its border with Mozambique. They remain little visited and without permanent habitation.

6. These myths are currently being researched by Jesse Sagawa, of the University of Fredericton, New Brunswick, who is a native of the Mulange district.

7. Haggard's cultivation of the myths of Mulange and its 'mystique' created a minor literary genre that was emulated by the late Laurens Van der Post in his controversial *Venture into the Interior*, a tale based on the latter's trip to Mulange in 1949.

8. Nigel Leask, *British Romantic Writers and the East: Anxieties and the Empire* (Cambridge: 1992); John MacKenzie, 'Edward Said and the Historian', *Nineteenth Century Contexts*, 18 (1994), pp. 9–25.

9. See Margery Perham, *Lugard: The Years of Adventure* (London: 1961).

10. Grove, *Green Imperialism*.

11. Richard Grove, 'Scottish Missionaries, Evangelical Discourses and the Origins of Conservation Thinking in Southern Africa, 1820–1900', *Journal of Southern African Studies*, 15/2 (1989), pp. 22–39.

12. See especially Thomas Pringle, *African Sketches* (London: 1834); Damien Shaw, 'The Life of Thomas Pringle', PhD thesis (Cambridge, St Edmunds College: 1996), and personal communication; J. R. Doyle, *Thomas Pringle* (London: 1972).

13. Pringle, *African Sketches*, p. 124.

14. Brown, *Pastoral Discourses*.

15. For details of Hooker's views and his attributions to Brown, see 'Forestry', in *Journal of Applied Sciences*, 1 (1872), pp. 221–3.

16. Simon Schama, *Landscape and Memory* (London: 1995); Tim Bonyhady, 'The Giant Killers', *Sydney Morning Herald*, 3 February 1996, p. 9; Tom Griffiths, *Secrets of the Forest* (Sydney: 1992), chapter 1 and p. 143.

17. J. C. Brown, 'Early Planting in Scotland', *Transactions of the Royal Scottish Arboricultural Society* (1881), p. 355.

18. *Life of Johnson* (London: 1787), ed. Hill, vol. 3, p. 59.

19. Among other references to Rousseau, see his *Origin of Language*, 2nd edn, lines 403 and 414 n., and *Antient Metaphysics*. Chapter 12 in the first volume of the former work is avowedly an attempt to solve 'M. Rousseau's great difficulty with respect to the invention of language'. In his preface to the *History of the Wild Girl*, Monboddo says that Rousseau is the 'only philosopher of our time' who has conceived the *magnum opus* of philosophy to be 'to enquire whether, by the improvement of our faculties, we have mended our condition

and become happier as well as wiser'. But, he adds, 'though Rousseau had the idea, none has executed it'.

20. For the connections with French practice, see George Hamilton and J. Balfour, *A Treatise on the Manner of Raising Fruit Trees; in a Letter from the Earl of Haddington, to which Are Added Two Memoirs; the One on Preserving Forests; the Other on the Culture of Forests Transl. from the French of M. de Buffon* (Edinburgh: 1761). Vilmorin, the famous French experimental forester, reported his reliance on the 1760 published work of Thomas Hadding-ton and particularly credited his work on the culture of *Pinus sylvestris* (the Scots pine): see Brown, *Pine Plantations on the Sand Wastes of France*.

21. The reafforestation of the Haddington district was widely remarked by contemporary com-mentators. For example, Mrs Piozzi, in her *Thraliana* journal, notes that 'for never were more elegantly disposed plantations scattered round a country, and both the Duke of Roxburgh's seat and Haddington have a vast quantity of ancient and respectable timber about them and very fine thick hedges adorn and shelter out roads': quoted in R. R. Reynolds, 'Mrs Piozzi's Scottish Journey', *Bulletin of the John Rylands University Library of Manchester*, 60 (1977), p. 116.

22. P. J. Venter, 'An Early Botanist and Conservationist at the Cape: The Reverend John Croumbie Brown, LL.D, F.R.G.S., F.L.S.', *Archives of the Cape Museum* (1951).

23. See Brown, *Forestry in the Mining Districts of the Ural Mountains in Eastern Russia* and *Forests and Forestry in Poland, Lithuania, the Ukraine and the Baltic Provinces of Russia*.

24. Brown, *The Truth and Truths of Christianity*; Brown, *Pastoral Discourses*.

25. Brown, *Pastoral Discourses*, p. 94.

26. Brown, *Pastoral Discourses*, p. 96.

27. Brown, *Pastoral Discourses*, pp. 97–8.

28. Brown, *Pastoral Discourses*, p. 108.

29. Brown, *The Hydrology of South Africa*, 'Preface'.

30. Personal record of J. C. Brown, King's College, Aberdeen, Archives.

31. Compare R. H. Grove, 'The East India Company, the Australians and the El Niño: Colonial Scientists and Analysis of the Mechanisms of Global Climate Change and Teleconnections between 1770 and 1930', *Working Papers in Economic History*, 182 (Canberra: 1995).

32. J. T. Arbousset and P. Daumas, *Narrative of an Exploratory Tour to the North-East of the Colony of the Cape of Good Hope* (ed. and trans. J. C. Brown).

33. See J. C. Brown, 'The International Forestry Exhibition in Edinburgh: An Argument for the Organization of a National School of Forestry', paper read before the British Association in Aberdeen (1885).

34. See also William Beinart, Chapter 6, above, and Shaun Milton, Chapter 13, below.

35. George Baden-Powell, 'New Homes for the Old Country', *Spectator*, quoted in Brown, *The Hydrology of South Africa*, p. 1.

36. The most notable of these was a treatise on 'Improved Forest Management in India and Burmah etc.'. Brown clearly decided to leave the complexities of Indian forest management to the likes of Cleghorn, and tended to avoid the subject of the sub-continent.

37. J. Wilson, 'On the Progressive Desiccation of the Basin of the River Orange in Southern Africa', *Proceedings of the Royal Geographical Society* (1865), pp. 106–9.

CHAPTER 10

Mawson of the Antarctic, Flynn of the Inland: progressive heroes on Australia's ecological frontiers

Brigid Hains

In early 1909 young Douglas Mawson, prematurely wrinkled by the ravages of wind and hunger, returned from Antarctica to a hero's welcome. Public enthusiasm lasted long enough for him to raise £40,000 for his own Antarctic expedition by the end of 1911. He was a charismatic and forceful leader and became an icon of Australian resourcefulness, bravery and individualism. He and his men experienced the environmental extreme of Antarctica, and their experiences illuminate the uneasy relationship between nature and society in early twentieth-century Australia.

Australians did not need to go as far as Antarctica to confront the ecological frontier. As Tom Griffiths has observed: 'for adventurous and scientific Australians of the early twentieth century, two frontiers beckoned: the white ice and the red heart, the far south and the immediate north'.[1] Scientists were not the only urbanites to be drawn to the 'red' or 'dead' heart of Australia. The life of John Flynn, creator of the Flying Doctor Service, provides a useful parallel to Mawson's story of survival in the Antarctic.

The two were utterly different: Mawson was a heroic explorer, élite scientist and technocrat; Flynn was a missionary, Church bureaucrat, anti-intellectual and a reformer of outback life. Yet they were both folk heroes of a society preoccupied and uneasy about its tenure in a strange and unyielding environment. Each in his own way represented the transcendence of environmental limitations, the triumph of the distinctively *human* spirit over nature. Their lives offer a window onto the negotiation between environment and Australian settler which reaches a crescendo in the early part of the twentieth century.

This chapter took shape in the context of conservation history concerned with the genesis of wilderness appreciation. By the late twentieth century, Antarctica had become the archetypal international wilderness, and I sought to tease out the 'antecedents' to this romance with the ice. Conservation tends to occupy the 'high moral ground',[2] and it seemed perhaps that too many environmental historians do likewise: the search for antecedents to contemporary sensibilities tends to conceal less savoury aspects of environmental consciousness.[3] Early twentieth-century currents of cultural development such as progressive conservation, bushwalking, nature writing and national parks share the same cultural space as less savoury issues such as eugenics, Royal Commissions into the birth rate, fears of racial degeneration in the tropics, and acrimonious controversy over the future of white settlement in the continental interior.

Conservation – and especially progressive conservation – is just one of the threads woven into a tapestry of intense preoccupation with the biological dimensions of society between the late nineteenth and the early twentieth century. Nancy Stepan has described the development of 'racial science' in this period as:

> a shift from a sense of man as primarily a social being, governed by social laws and standing apart from nature, to a sense of man as primarily a biological being, embedded in nature and governed by biological laws ... In short a shift had occurred in which culture and the social behaviour of man became epiphenomena of biology.[4]

Eugenics and racialism are two obvious examples of this 'biologization' of society,[5] but I want to link these with other cultural preoccupations into a broader picture of a society in ongoing negotiation with its environment. Two contradictory trends in contemporary thought add urgency to this inquiry. On the one hand, contemporary environmentalism is reviving biological determinism in relation to social policy (including immigration).[6] On the other hand, postwar liberal humanism has vigorously repudiated biological determinism.[7] The rejection of biological determinism, along with a suspicion of science, is equally characteristic of the cultural theories which seek to replace liberal humanism in the academy. In this paradigm, biological determinism is most often read as a culturally determined product of oppressive knowledge structures which seek to limit freedoms of the individual. It seems timely, then, to review the historical role of biological determinism at the beginning of our own century.

Evolutionary theory is one of the richest lodes for the cultural historian of the late nineteenth and early twentieth centuries to mine. It is not surprising, then, that both Mawson and Flynn were preoccupied with evolution and regression on the environmental edge. Even more interesting are the striking parallels between their responses to these possibilities – responses which reflect the centrality of progressive thought to environmental perceptions and practices in this period.[8]

Mawson had first gone to the Antarctic in 1907 with Shackleton. He was a promising young geologist and had already been on scientific expeditions to the Pacific.[9] Shackleton and Mawson planned another expedition together, but Shackleton withdrew, whereupon Mawson ingeniously reconstructed the expedition in a patriotic fashion as the Australasian Antarctic Expedition.[10] The scientific establishment was enthusiastic in its support – many scientists had been agitating for an Australian expedition since the late 1880s.[11] In 1911 two continental bases and one sub-Antarctic base were established, but the main base, under Mawson's command, was dogged by appalling weather which made outdoor work almost impossible. During the summer, Mawson and two others embarked on an ambitious inland sledging trip. Mawson alone survived, arriving back at base after a gruelling 300-mile solo trek only to find that the relief ship had left that same morning. Mawson and the six others who had been left behind by the ship to search for the missing party bunkered down for a second, much grimmer, year at the base. They finally returned in 1914, to a world on the brink of war.

It was a convention amongst scientists to regard an expedition to Antarctica as a journey in time as much as space. On Scott's last expedition, this paradigm prompted an arduous winter trek to collect an egg from an emperor penguin. Believing the penguin to be the most primitive bird, Scott's scientists hoped that its embryology would reveal the evolutionary link between reptiles and birds.[12] Like Australia, Antarctica was seen to be a museum continent, inhabited by 'creatures, often crude and quaint, that have elsewhere passed away and given place to higher forms'.[13] In the diaries of Mawson's expeditioners, the scientific assumption of primitiveness elided into anthropomorphic images of arrested development, such as the stupidity of the crab-eater seals when faced with threatening dogs and humans.[14] The predatory leopard seal was more atavistic: a 'weird ... awesome ... monster ... the expression ... was very horrible'.[15] This moral primitivism owes as much to the conventions of Victorian gothic as to a scientific model of the natural world.[16] But primeval could also mean innocent: snow petrels could be picked up from the nests, 'angelic little creatures ... this is the sort of thing one reads of in books where an explorer discovers a rare bird which has never looked on man'.[17] Like the positive designation of wilderness as 'primitive' by bushwalkers, the primeval Antarctic could be a retreat from the decadence of urban life.[18]

Antarctic geology, too, seemed to be in the throes of evolution. Like Australia, the Antarctic continent invited contradictory images of antiquity and newness. The geology appeared viscous: a snow-covered ridge 'like the back of some gigantic, plated saurian creeping under the snow'.[19] Despite this dynamic topography, the land could appear drained of vitality, 'weird' and desolate: 'All this land is Death. Here it was both death and decay – it might have been the graveyard of centuries it looked so old.'[20]

No wonder, then, that the expeditioners' speculations about the influence of the Antarctic environment on their physical and moral selves were deeply ambivalent. On the one hand, the frontier exerted a therapeutic stress on mind and body: 'cold work but we enjoyed it'.[21] 'A Perfect day in Antarctica is Heaven', declared another; for one participant, the expedition was a cure for his 'nerves': 'speaking absolutely truthfully I have never in my life felt as well, this cold is bracing, seems to fill you with vitality ... I am so contented and happy it is great to realise an ambition & I have always wanted to get into the new lands.'[22] The vitality of the frontier life was explicitly allied to the imperial venture. Mawson has the final word: 'It seems to me that man and his brain have evolved as air breathing animals, and in temperate and cold climates because of the vicissitudes there.'[23]

Yet the antithesis of this environmental optimism was the fear of degeneration. The regression of the expeditioners to an uncivilized state was a standing joke. By July 1912 Harrisson could remark on 'wearing the same clothes as last April'; on sledge trips they resorted to 'cave-dwelling'; and Dovers remarked on the fragility of the bonds of civil behaviour when he wrote home: 'I will talk learnedly of ice blink and water skies etc when I get back and Kiddo will have to take me in hand and teach me modern conversation as I will only be able to talk of ice and snow, also you will have to run me into the bath every morning and see I wash ... '.[24] In fact, they exhibited the classic signs of 'going native': the loss of habits of hygiene and gentility; and the abandonment of civilized conversation for the language of natural phenomena.[25] Their mental world was exchanged for an imagination suffused with, and limited to, the natural environment. At one point, Antarctica was described as 'no white man's land'.[26] And all this was in spite of the absence of 'natives' to emulate. The contrast between civilization and the wild was graphically demonstrated to Mawson when he became disoriented in a blizzard, although less than 300 metres from the hut:

> We return ... after a great struggle to reach the Hut. Blizzard strong, dense drift and dark ... Have to do on hands and knees feeling one's way with ice-axe ... walk on top of house without knowing it. Feel with ice-axe ... At last after probing in every direction feel space below and plunge in. I had thought of the advisability of camping on the rocks – now everything's different and I forget the danger of a few minutes ago and go on with notes on ice and its structures. No light from the Hut it is difficult to tell when one is on top of it. Outside one is in touch with the sternest of Nature – one might be a lone soul standing in Precambrian times or on Mars – all is desolation and life in the durest [sic]. Life opens up to one as it must to the savage. Inside the Hut all is 20th century civilization. What a contrast.[27]

'Life opens up' to Mawson in this confrontation with nature, but the expansive moment is vitiated by its desolation. That frontier life was both ennobling and regressive was a common theme of popular frontier literature, a genre which was itself a significant feature of life on the expedition. (Mawson, for example, read aloud to the men from the works of Robert Service.) The concern for vitality, the strenuous life and the cleansing effects of the frontier are interwoven in this literature, and their connection with regression found popular expression in the tremendous success of Jack London's *Call of the Wild*, which sold over a million and a half copies between 1900 and 1933.[28] Note the contrast between man in nature – stranded in the brutal uncertainty and darkness of the blizzard – with the domestic interior of the hut, where routine, light, and hierarchy all played a role in maintaining civilization and *excluding* unwanted elements of nature. When Mawson returns to the hut, he continues with his notes on ice structures: the same substance which threatened to overwhelm him moments earlier is contained and subjected to rational analysis.

Flynn, too, was equivocal about the ecological edge, the internal frontier of Australia's 'Inland'. We know this through the legacy of the rich environmental rhetoric which we find in both his personal papers and in his journalism, especially for *The Inlander,* a mission magazine with a circulation of 5,000– 10,000 between 1912 and 1929 for which Flynn was editor and principal contributor.[29] Flynn was an unhappy theological student, just scraping through his formal education as a Presbyterian minister.[30] His real love was his Home Mission work in the remoter rural areas of his native Victoria. He rode trains and horses to reach the timber-cutters of the Otway Ranges and the Snowy River and the shearers of the Western District. He lectured with lantern-slides and was an enthusiastic amateur photographer. By 1910, still a student although nearly 30, he knew enough to produce *The Bushman's Companion*, a booklet with advice on first aid, burial services, and exhortations to positive thinking. The mixture of technical advice and pragmatic moralizing was to become characteristic of Flynn's mature style. Ten thousand of these handbooks were distributed free to shearers and others in its first edition, and it was republished annually for some years.[31]

Flynn was an ambitious young man who wanted a high-profile urban ministry, but he was offered the post of minister at Beltana in the arid north of South Australia. Here he formulated his vision of a comprehensive inland mission to serve all of remote Australia. He manoeuvred himself into the job of compiling a Church report on the Northern Territory, launching his lifetime career as Superintendent of the new Australian Inland Mission (AIM). As early as 1910 Flynn expressed a classic Australian ambivalence in *The Bushman's Companion*: 'After all it is not necessary to have met face to face to feel a sense of companionship. We have a mutual love of the bush, and along with that, perhaps, a certain dread of it.'[32] By 1915 Flynn had consolidated his vision of a nation dependent on the pioneer fringe for its vitality and future development. Bush dwellers were respectable – 'many are highly educated and the majority are keen active men'[33] – and resilient in the face of the vicissitudes of an unpredictable environment and its capricious weather. In the wake of the Anzac landing, Flynn's adulation of the bushman increased, and he featured Australian buck-jumpers in Egypt in *The Inlander*.[34] More dramatically, Flynn started to argue for the evolutionary benefit of the bush – particularly the semi-arid pastoral lands: 'Cattle are raised in valleys among these mountains, and no finer horses are to be found anywhere in Australia than those bred here – beautiful of body and limb, timid and shy, but wonderful in stamina ... Stockmen are akin to the horses in endurance and courage.'[35] In the 1920s Flynn's descriptions of the bushmen took on a decidedly eugenic edge:

If thousands – it is a matter of thousands – of our most virile and adventurous pioneers, comprising our A1 human stock, are allowed to remain celibate, sinking under the scythe of time without trace: and if the lands to Centre, and North, and West are thus allowed to remain practically void of real family life, that will indeed be the Funeral of Australia.[36]

It is unclear here whether the pioneers were self-selected for initiative and endurance or whether these characteristics were an outcome of their encounter with the environment. Flynn thus reflected a popular conflation of hereditary and environmental factors in evolution: a mix of strict hereditarian evolution, Lamarckism and social environmentalism.

Flynn was now relinquishing his early advocacy of closer settlement, substituting a vision of a permanent and positive frontier. In an article entitled 'Desert and Destiny', he claimed that, just as the marginal lands had encouraged the breeding of better agricultural products such as hard wheat and fine wool, so they provided an alternative to urban life. City folk were a nation of 'straphangers' whose lives revolved around 'the models of frocks and cars and songs and sentiments of every country but their own'. As a remedy, God had provided the desert fringes, where 'the danger-waves of bloated social organisms curl, break and recede' and where 'choice spirits' could receive an 'opportunity to breathe into the body, and mind and spirit, deep draughts of virgin air'.[37] Flynn's ecological frontier would breed the best of the coming generation, he argued:

> Who would place any limit on the possibilities in the life of a child born of adventurous parentage, and reared on natural foods: trained in vastly spacious playing grounds, where wonderful silence reigns; honoured in toil and in sport by comradeship with the hardiest of ADULT mates; and all the while provided with a liberal share of the World's best literature? [38]

Furthermore, Flynn foresaw the day when the outback would be a popular holiday spot for people of modest means whose lives would be refreshed and who would return to the city providing 'that leaven by which our national loaves may be saved from heaviness'. It must have seemed as far-fetched at the time as Mawson's idea that Antarctica would in future be 'the scene of summer pleasure cruises from Australia and New Zealand'.[39]

Flynn's vision was imbued with a deeply racialist ethos of environmental influence. Anglo-Saxons, he argued, like Jack London and many other contemporaries, were natural pioneers – indeed, they needed to expand to maintain their vitality: 'Actually the so-called Aussie spirit was induced originally in Pommies, thanks to lonely environments everywhere, which immediately called forth inherent powers that had long been lying dormant in the race which transplanted itself into this new land.'[40] This was a convenient way of explaining why the positive environmental influence of the bush had manifestly failed for the Aborigines. Flynn argued that their lack of efficiency had precluded them from 'effective occupation' of the land.[41] Racial mixture was a deep concern for Flynn: 'destiny is worked out, humanly speaking', he said, 'according to natural features'.[42] The Aborigines seemed a signal failure in the evolutionary stakes: 'The gulf between the Australian Aborigine and you is so deep and wide – as you yourself are glad to think – that several stages of evolution might fairly be allowed before demanding much attainment on the other side.'[43]

The civility of the white bush settlers could be polluted by Aboriginal customs or, worse, by Aboriginal genes, with the bushman 'living like a black-fellow',[44] producing 'ignorant half-castes'.[45] An early stimulus in Flynn's work was a letter from a white woman in Darwin deploring the settler practice of 'interfering with lubras' as both damaging to Aboriginal women and, more importantly, degrading for whites: 'know that drink, drugs and lubras are responsible for nine out of ten hospital cases & responsible for seven deaths out of every ten'.[46] This association between public health and racial purity was to be a keystone in Flynn's solutions to the problem of inland settlement. Another of his correspondents worried about white children who were subject to the same stultifying environment as the Aborigine:

> I have been thinking over a conversation I had lately with a married man, the father of a growing family, he told me quite seriously that Australian children were reverting to the blacks, not in colour, but in manner and habits; that they had not the energy, enterprise and perseverance of their fore-fathers ... the seeming retrogression is due to lack of intellectual exercise and religious instruction.[47]

In other words, the children of the outback were reduced to the same state as Mawson, lost outside the hut in a blizzard: excluded from civilization, true domesticity and the intellect and immured in nature unmediated, they were like savages – and Flynn had no time for 'cheap talk about the joys of primitive conditions of life'.[48] While he believed in the ennobling qualities of the pioneering life, he was anxious about the possible slide into degradation. The call of the wild was energizing but dangerous.

It is hardly surprising, then, that both Mawson and Flynn, despite the gulf between them, resorted to the same set of progressive ideals to shape their policies as leaders and innovators. Mawson's Antarctic base – a kind of colony in microcosm – and Flynn's 'Inland' Australia both constituted ambitious social experiments in their own way, experiments in which the physical and moral health of the participants were intimately bound up with their environmental strategies. Both shared the dream of a highly functional society in which efficiency and progress were key criteria for success. Moreover, this vision was not limited to society in the strict sense but was extended to refer to the interaction between society and the natural environment.

The progressive conservation movement is well known to environmental historians.[49] I want to argue here, however, that the impact of progressive ideology on environmental sensibilities goes far beyond the conservation movement. One of the most interesting aspects of progressivism is the elevation of efficiency and scientific expertise to a kind of religious enthusiasm; this was certainly central to both Mawson and Flynn. The distinctive tension at the heart of progressivism was, however, the desire to use the means of highly industrialized culture (technology, science, expertise and efficient management) to counter the perceived effects of industrialization (social atomization, destruction of natural resources, and ill-health). If there is one concept which best describes progressive ideals, it is 'efficiency', and both Mawson and Flynn were 'efficiency men', although what it actually meant to them requires some analysis.

Mawson's Australasian Antarctic Expedition was imbued with utilitarian goals from the first stages of planning. Prospectuses to promote the expedition did not engage in the romantic rhetoric of exploration, but emphasized the scientific credentials, efficiency and usefulness of Antarctic exploration.[50] The conduct and experience of the expedition was also imbued with scientific utilitarian ideals. It was the first Antarctic expedition to use wireless, and this became a symbol for the struggle of technology to overcome remoteness and environmental hostility. Two wireless masts, each nearly forty metres high, were needed to sustain communication with shipping and mainland Australia. Mawson was particularly committed to the wireless, driving the men on even when the struggle seemed hopeless and the masts had been blown down repeatedly in the constant blizzards. It was not until the second year, when the expedition members were stranded, that the wireless was used successfully; it then became a vital boost to morale. It was 'quite like a morning paper to hear the contents [of overnight transmissions] at breakfast time'.[51] Communication was lost for a time in winter, but when it was re-established, the expeditioners were delighted: 'my pulse is quite normal tonight and I don't feel the least bit excited but we are secretly exultant at having won the battle with the elements and having shown the world the same ... so we embark on what seems a new era – returning spring and news from Australia'.[52] The battle against the elements is an internal as well as an external one; involving the unreliable biology of the self, which may succumb to nervous excitement, as well as the hostile weather and natural conditions. By means of the wireless, the expedition was knitted back into civilization and, of course, the ecological frontier was domesticated through the triumph of a quintessential modern technology – the technology of long-distance communication.

For Flynn, this ideal of a technologically unified society was even more powerful. Flynn's initial plan for the AIM was centred on the 'patrol padre': a nomadic minister who visited remote settlements and stations. Within a decade, however, it became clear that the job was simply too onerous: the harsh climate, the continual travelling in appalling conditions, the rough society of the mining camps and pastoral settlements all took their toll. Simultaneously, Flynn was attempting to develop technological solutions to the dilemma of a scattered population in a harsh environment. Like Mawson, he was gripped by the vision of the technology of communication and transport defeating distance and remoteness, despite the differences between the two 'frontiers'. Notwithstanding his enthusiasm for peopling the outback, Flynn gradually accepted that 'a large portion of Australia must be sparsely occupied by human beings'.[53] He began to rework his ambitions to minister to a continent with a *permanent* frontier, but a frontier which – like Mawson's base in Antarctica in its own small way – was knitted into civilization by remote control. Railways were 'steel highways which link up British peoples as the Roman road united that Empire 2,000 years ago' – agents and markers of civilization, but wireless and aviation were the most important elements in his strategy.[54]

Wireless was a prelude and a precondition to Flynn's greater dream of the flying doctor: the provision of communication and transport across the 'Inland', as well as the psychological comfort of a 'mantle of safety'. Public health was close to the progressive heart, of course, especially the reform of conditions of work and home life. Flynn's concern for public health in the remote settlements extended from the provision of emergency care by nursing homes and the flying doctor to the reform of building methods which would make outback homes sanctuaries against the severity of the northern climate.[55] A project which engaged much of his attention, for example, was the building of a hospital in Alice Springs which used ingenious methods of evaporative air cooling.[56] The building was promoted as being highly efficient, but Flynn made it clear that efficiency was something quite different from economy. As a review of the British Civil Service had stated in 1900: 'Nothing is so cheap as efficiency: nothing is so inefficient as cheapness. Substitute for the Treasury control which seeks cheapness the Administrative control which seeks efficiency, and the nation will be the gainer, even pecuniarily.'[57] Flynn himself recommended 'wisely reckless expenditure' on frontier development, since greater issues than financial prudence were at stake.[58] Efficiency, health and social cohesion were bound up in Flynn's mind with the health of settler society as a whole, particularly with its racial vitality:

> A word may be in season re the general question of buildings for Inland ... Our pioneers have gone out into their battle ill-equipped as a rule, and at the outset comfort has necessarily been ignored.
>
> But what is 'comfort'? Is it luxury, merely? Surely it rather signifies *conditions favouring highest human efficiency.*
>
> Look for instance, at the brave bush wife who cheerfully cooks her family's dinner at a stove in a lean-to off the low living-room ... This citizen never complains, even to herself: but because of the years of such defiance of 'comfort' by a mother, many a child is permanently weakened in vitality. Also the home is robbed of some brightness ... And the nurse who loves making the best of inadequate equipment: she pays out of her own spirit more than is good for the sake of the patients who need her utmost vitality in service for them.
>
> ... [the goal must be] to lessen the nerve-strain of frontier life ... It is hardly necessary to moralize at length about the influence of good buildings – apart from their intrinsic value ... Buildings talk eloquently in their silent

way. The bushman who has for years been denying himself all comforts, consciously or unconsciously doing his bit to build Australia Tomorrow, does not always forget that he is 'living like a blackfellow'. Time comes when he bitterly remembers that fact and still more bitterly tells himself that no-one cares whether he lives or dies in his isolation.

Imagine such a man, being injured, brought from afar into a Nursing Home fitted up in its small way according to the very highest standards of efficiency and quiet taste – both for hospitals and homes ... he revises some of his thinking: he goes out knowing that the Nursing Home (The more it has cost the better now!) is the price some distant, unseen fellow-citizens have placed on his life.[59]

Efficiency was yoked to virtue, as it was for American scientific progressives who believed that 'the efficient person was moral because he served society, and society would become more moral as it became more efficient'.[60]

To Flynn, morale was a vital element in pioneer confidence; it was not until the *morale* of the frontier population improved that the *morality* of the population would improve also. The progressive ideal of a streamlined, efficient, virtuous and socially cohesive private life flowered in the ideal of progressive womanhood.[61] One of Flynn's lifelong campaigns was to encourage women to settle in the bush in order to create a truly civil frontier society: there was 'no greater need nor greater hope'.[62] Without women, the masculine frontier was a dangerous place for the morale of the nation. Without women, men on the frontier were vulnerable to becoming what Mawson experienced momentarily in the blizzard: savages submerged in nature and submerged too often in the company of savage people. To Flynn, women represented order, domestic arts, health and the maintenance of cultural and social standards in the wilderness. Perhaps the most telling demonstration of this was the station garden – a sure sign of the presence of a white woman in the homestead and a sign, too, of the arrival of a health-giving positive environment, including the nutritional benefits of fresh vegetables as well as the more subtle impact of an ordered and comfortable garden, like the impact of the well-built nursing home.

In a very different context, Mawson also recognized the importance of 'feminine' virtues of craft, care and efficiency above romantic risk-taking. Every pound of weight was crucial in Antarctic sledging, and the refinement of clothing, camp methods and equipment could mean the difference between survival and gruesome death. Mawson set a strict regime for winter work in his hut, much stricter than that of the other base. Shirking was Mawson's pet hate. He once recorded in his diary with disgust that an expeditioner who had been sent inside to make lunch on one of the few fine days was found in his bunk reading Roosevelt's *The Strenuous Life*![63] Mawson was impatient with ill-health, writing of the same man: 'Close has been laid up with a severe gum boil. He should never have come to the Antarctic with such remnants of teeth.'[64] Like Flynn, Mawson conflated health, efficiency and morality. But he also expected some of the civilizing virtues which Flynn identified as the feminine contribution to the well-being of the outback: cooking decent, nutritious meals with food that was as fresh as possible; looking after equipment; keeping an orderly 'household'; washing clothes; and repairing sledging equipment. Such tasks made the men appreciate their mothers, who did the 'finnicking [sic] things in life'.[65] Mawson was equally impatient with an expeditioner who valued 'derring-do' over efficient preparedness: 'The vista [in a blizzard] is a chaos. Webb dived off in this for the Magnetic Hut – in an instant he was lost to sight ... I wish he would look after his clothes better. For though he stands the conditions very well, he is affected by over much bravado; at least I trust it will not bring him any harm ...'.[66]

Mawson and Flynn were dealing with a highly moralized landscape in their encounters with the ecological frontier. Each constructed strategies for dealing with these harsh environments which sought to control and order the relationship between settlers and nature. Each was careful to preserve the invigorating potential of the nexus between new lands and the Australian character. Apart from the rhetoric of race, evolution and regression, it is clear that progressivism provides the framework for much of this negotiation, with its peculiar blend of nostalgia for the organic, land-holding society allied to an intense enthusiasm for technology and scientific expertise. Despite the different social and natural environments in which each leader operated, they sought parallel solutions to the challenges of nature: solutions which sought to control the moral as well as the physical relation between settler and frontier, to control the inner world of domestic life as much as the public or outer world of exploration, farming and mining. This similarity, together with the fame and popularity of both men, is a tribute to the pervasiveness of the progressive ideal in Australian society in this period.

Notes

1. Tom Griffiths, *Hunters and Collectors: The Antiquarian Imagination in Australia* (Melbourne and Cambridge: 1996), pp. 178–9.
2. Libby Robin, 'Of Desert and Watershed: The Rise of Ecological Consciousness in Victoria, Australia', in Michael Shortland (ed.), *Science and Nature: Essays in the History of the Environmental Sciences* (Oxford: 1993), p. 115.
3. A recent example is Colin Michael Hall's *From Wasteland to World Heritage* (Carlton: 1992), which uses a whiggish model to explain the development of wilderness consciousness in Australia.
4. Nancy Stepan, *The Idea of Race in Science: Great Britain 1800–1960* (London: 1982), p. 4.
5. Andrew Markus, *Australian Race Relations* (St Leonards: 1994), p. 1.
6. A particularly sophisticated form of this argument is in Timothy Fritjof Flannery, *The Future Eaters* (Sydney: 1994; repr. 1995), especially chapter 33.
7. Carl Degler, *In Search of Human Nature* (New York: 1991); Elazar Barakan, *The Retreat of Scientific Racism* (Cambridge: 1992).
8. A concern for an appropriate acknowledgement of the contribution of progressive ideas to environmental management is evinced in J. M. Powell, *An Historical Geography of Modern Australia* (Cambridge: 1988).

9. Mawson's career is outlined in the introduction to F. and E. Jacka (eds), *Mawson's Antarctic Diaries* (Sydney: 1988).

10. Mawson to Shackleton, [?] 1910 (Mawson Papers, University of Adelaide [UA], 11AAE).

11. Lynette Cole, *Proposals for the First Australian Antarctic Expedition* (Clayton: 1990).

12. Apsley Cherry-Garrard, *The Worst Journey in the World* (Harmondsworth: 1948), pp. 34, 256.

13. Baldwin Spencer, quoted in D. J. Mulvaney and J. H. Calaby, *'So Much that Is New': Baldwin Spencer 1860–1929, a Biography* (Carlton: 1985); the quote refers to Australia.

14. A. McLean, Diary, 8 March 1912 (Mitchell Library Manuscripts Collection [ML]).

15. McLean Diary, 3 April 1912.

16. It may also reflect an older convention in European thought which characterized the predator as a villain: see Donald Worster, *Nature's Economy: A History of Ecological Ideas*, 2nd edn (Cambridge: 1994), chapter 13.

17. McLean Diary, 12 November 1913. See also the discussion of 'innocent' game animals in Africa in John M. MacKenzie, *The Empire of Nature: Hunting, Conservation and British Imperialism* (Manchester: 1988), pp. 137–8.

18. M. Harper, 'The Battle for the Bush: Bushwalking versus Hiking between the Wars', *Journal of Australian Studies*, 45 (1995), pp. 41–52.

19. C. Harrisson, Antarctic Diary, 6 November 1912 (ML).

20. Harrisson Diary, 2 October 1912.

21. McLean Diary, 16 November 1912.

22. C. Laseron Diary, 27 April 1912; R. Dovers Letters, 7 January 1912 (both ML).

23. Mawson Diary, 11 May 1912 (in F. and E. Jacka (eds), *Mawson's Antarctic Diaries*).

24. Harrisson Diary, 11 July 1912; M. H. Moyes Antarctic Diary, 7 September 1912 (ML); Dovers Letters, 2 February 1912.

25. 'Going native' was the theme of 'captivity narratives' in both Australia and America: see R. Dixon, *Writing the Colonial Adventure: Race, Gender and Nation in Anglo-Australian Popular Fiction, 1875–1914* (Cambridge: 1995), chapter 3.

26. Harrisson Diary, 29 March 1912.

27. Mawson Diary, 9 April 1912.

28. P. J. Schmitt, *Back to Nature: The Arcadian Myth in Urban America* (New York: 1969), p. 125; see also T. Gossett, 'Literary Naturalism and Race', in *Race: The History of an Idea in America* (Dallas: 1963), pp. 198–227.

29. Circulation of *The Inlander* in Australian Inland Mission papers (National Library of Australia (NLA), MS 5774, Box 11, Folder 2).

30. This summary of Flynn's life is based on his correspondence in Flynn Papers (NLA, MS 3288, series 1).

31. Flynn Papers, Series 1.

32. J. Flynn, *A Bushman's Companion*, 1916 edition (Flynn Papers, Box 15).

33. *The Inlander*, 2/1 (1915), p. 12.

34. *The Inlander*, 5/2 (1919), p. 60.

35. *The Inlander*, 3/1 (1916), p. 6.

36. *The Inlander*, 7/1 (1922), p. 7.

37. *The Inlander*, September (1924), pp. 90-1. Flynn's 'choice spirits' seem akin to Sydney's élite interwar bushwalkers: see Harper, 'The Battle for the Bush'.

38. *The Inlander*, September (1924), p. 93.

39. Promotional material in Mawson papers (11AAE).

40. *The Inlander*, September (1924), p. 92; see also Gossett, *Race*, pp. 198–227.

41. *The Inlander*, September (1924), pp. 21–2.

42. *The Inlander*, 4/1 (1917).

43. *The Inlander*, 2/1 (1915), p. 26.

44. *The Inlander*, 6/1 (1920), p. 67.

45. *The Inlander*, 7/1 (1922), p. 193.

46. Jessie Litchfield to J. Flynn, 15 May 1911, and Jessie Litchfield to Mrs Kelly (Flynn Papers, Series 1, Folder 1).

47. D. McKellar to J. Flynn, 26 May 1910 (Flynn Papers, Series 1, Folder 1).

48. *The Inlander*, 2/1 (1915), p. 50.

49. The classic work is Samuel P. Hays, *Conservation and the Gospel of Efficiency: The Progressive Conservation Movement 1890–1920* (Cambridge, Mass.: 1959).

50. In Mawson Papers (11AAE); see also Stephen J. Pyne, *The Ice: A Journey to Antarctica* (Iowa City: 1986), pp. 168ff.

51. McLean Diary, 9 March 1913.

52. McLean Diary, 5 August 1913.

53. *The Inlander*, September (1924).

54. *The Inlander*, 4/2 (1918).

55. He was particularly concerned about the effects of the climate on white women: see *The Inlander*, 6/1 (1920), p. 48.

56. *The Inlander*, 6/1 (1920), p. 51.

57. Lyttleton Gell, '1900', in Geoffrey Searle (ed.), *The Quest for National Efficiency: A Study in British Politics and Political Thought 1899–1914* (Oxford: 1971), p. 70.

58. *The Inlander*, 1/3 (1914), p. 111.

59. *The Inlander*, 6/1 (1920), pp. 65–6.

60. David Danbom, *'The World of Hope': Progressives and the Struggle for an Ethical Public Life* (Philadelphia: 1987), p. 121.

61. K. M. Reiger, *The Disenchantment of the Home: Modernizing the Australian Family 1880–1940* (Melbourne: 1985), especially the introduction.

62. *The Inlander*, 5/1 (1918), p. 26.

63. Mawson Diary, 1–6 April 1913.

64. Mawson Diary, 3 July 1912.

65. Mawson Diary, 13 April 1912.

66. Mawson Diary, 14 May 1912.

PART 4

Economy and Ecology

CHAPTER 11

Ecology, imperialism and deforestation

Michael Williams

Ecology and imperialism combined

The publication of Alfred Crosby's *Ecological Imperialism* in 1986 brought together for the first time two seemingly contrasting words and concepts.[1] 'Ecological' describes the relations of organisms with one another and their surroundings; it is generally considered desirable, good, and laudable. 'Imperialism', on the other hand, describes the supreme authority of one group over another and the acquisition and control of dependent territories, resources or peoples; it is generally considered disagreeable, bad, and derogatory.

The novel twist of Crosby's new combination of 'ecological imperialism', however, was that it not only depicted Europeans creating the neo-Europes of, in particular, North America, South America, Australia, and New Zealand, but it also envisaged their movement out of Europe as unleashing an unconscious and unpremeditated ecological/biological imperialism as their crops, weeds, germs and pests accomplished dramatic demographic takeovers.[2] Thus, it was the organisms which were imperialistic, because of their evolutionary ability to reign supreme over competitors and establish ecological dominance, rather than the people, the political systems or technological advances.

Whether or not one accepts the specific deterministic emphasis of evolutionary superiority in this thesis, the term 'ecological imperialism' has more recently become equated with any change in the ecology of a territory as a result of the takeover and/or penetration of one group by another. Some of these changes are certainly unpremeditated, but many are the result of conscious policies that can be traced to the purposeful imposition of new economic, political and social arrangements[3] and to the application of techniques – the 'tools' and 'tentacles' of empire as Headrick has called them – that were a part of the larger imperial design of past times.[4] Either way, however, the combination of 'ecology' and 'imperialism' to produce 'ecological imperialism', or the 'imperial conquest of ecology', has allowed an entirely new construct to emerge which has offered a prospectus or framework with which to illuminate certain dimensions of a new kind of history, vast in spatial scale and temporal scope, and addressing many of the concerns of a more environmentally conscious age.[5]

Nowhere is the appropriateness of this new approach more evident than with respect to the global forests, arguably the most desirable and useful of all the major natural environments of the world and, therefore, the most vulnerable to disturbance. While it did not need modern imperialism for the forest to disappear – it had been happening for millennia as populations increased in all societies, both eastern and western – the empire certainly accelerated the process. The impact of humans clearing the forest to create new land and to use the timber was all-pervading. 'With the disappearance of the forest,' wrote George Perkins Marsh in 1864 in his celebrated work, *Man and Nature*, 'all is changed.'[6] He marshalled an impressive array of material to show how the very physical environment itself was affected, its vegetation, hydrology, soils, rainfall, temperature, to say nothing of the transformed landscapes, economies and societies that ensued. In the measurement and elucidation of the human disturbance, Marsh thought that Australia would provide 'greater facilities and stronger motives' than would any other area of recent settlement. It was 'the country from which we have a right to expect the fullest elucidation of these difficult and disputable problems', he wrote, because its settlement was late and did not start until 'the physical sciences had become matters of almost universal attention'.[7] But it was not to be so. Science, in the sense of enquiry about the way in which the environment worked, did not flourish in Australia to the degree that it did in the USA, India or France. Moreover, the forest was such a relatively small part of the continent. Trees of any sort (including mallee scrub and savanna) covered about 238 million hectares, or a quarter of the continent, of which only about 30 million hectares (3–4 per cent) could be considered forest in the commonly accepted sense of the word and, therefore, suitable for commercial extraction.[8] In the Australian psyche, the 'outback' of plains, deserts and aridity loomed far larger than the forests.

It is suggested, therefore, that the Australian experience of deforestation should be considered in the wider, comparative context of the countries of the Pacific Rim and adjacent Indian Ocean which provide contrasting experiences of the ecological impact of imperialism through deforestation, as well as some fascinating parallels and commonalities.[9] Australia and North America are representative of what Donald W. Meinig has called the 'settler empires', a kind of imperial mutation from global trade to the colonization of new ground by permanent settlers. India provides the example *par excellence* of a 'nationalistic empire', built on the beginnings of a trading 'sea empire' but, ultimately, the most overtly imperialistic of all.[10] China and, to a lesser extent, Japan are correctives to the idea that all imperial designs of one group over another have to be 'western' or equated with 'colonialism', as in the customary example of Europe's predatory spread. Imperialism was not a specifically European process and it could be autonomous: in other words, it could be eastern and intramural.

The pace and locale of deforestation

The primary reason for forest destruction has always been the creation of new land in order to grow food. That is not to say that cutting for constructional timber was insignificant, or that the supply of fuel for heating, cooking and smelting was of little consequence. But these drains on the forests, particularly for fuel wood, were often the incidental accompaniments and outcomes of the larger enterprise of land clearing. Generally speaking, deforestation bore a rough relationship to population numbers and commercial activity; as the population of the countries around the Pacific and Indian Oceans more than doubled, from 450 million in 1700 to 1,719 million in 1920, so the demand for agricultural land (and for fuel) increased accordingly.

Table 11.1. *Cropland and land cover change in Pacific Rim continents, 1700, 1850 and 1920 (million hectares)*

	Cropland			Cropland Change		Forest	Change	Grassland	Change
				1700–1850	1850–1920	1700–1850	1850–1920	1700–1850	1850–1920
	1700	1850	1920						
North America	5	50	179	47	129	−45	−27	−1	−103
Australia and New Zealand	5	6	19	1	13	0	−6	−1	−8
Latin America	7	18	45	11	27	−25	−51	+13	+25
China	29	75	95	46	20	−39	−17	−1	−3
South Asia	53	71	98	18	27	−18	−28	0	+1
South-east Asia	4	7	21	3	14	−1	−5	−2	−9
TOTAL	103	227	457	126	230	−128	−134	8	−97

Sources: John F. Richards, 'Land Transformation', and P. Demeny, 'Population', both in B. L. Turner, II, *et al.* (eds), *The Earth as Transformed by Human Action* (New York: 1990), pp. 164, 41–54 respectively; Michael Williams, *The Americans and their Forests* (New York: 1989), pp. 357–61.

However, it is difficult to know exactly how much forest was felled to create new cropland. One calculation is presented in Table 11.1, which reconstructs the distribution of land use for 1700, 1850 and 1920.[11] The amount of cropland increased by 126 million hectares between 1700 and 1850 and a further 230 million hectares over the next seventy years. The forest decreased by 128 million hectares and 134 million hectares respectively during the same periods, a total decline of 262 million hectares. Before 1850, nearly all of the cropland came out of forest and woodland areas; after 1850, the proportion dropped to 58.5 per cent. From then on, far greater amounts of cropland were created in the open grasslands, generally easier to cultivate, particularly in the United States. There is every reason to think, however, that these figures of total forest conversion are conservative, as a more detailed study of North America suggests that the amount of 'improved land' – that is, cleared forest land, which was always more than cropland alone – rose by 92.2 million hectares after 1850, not the mere 27 million hectares shown in the table.[12]

The United States: settler empire

The neo-European world of North America extolled the virtues of agriculture, freehold tenure, dispersed settlement, and personal and political freedom leading to a rapid and successful expansion of settlement. Rarely was a tally kept of the amount of forest cleared, because, amongst other things, clearing was regarded as the first step in the 'natural' process of 'improvement', so obvious and commonplace as to be barely worth comment or record. To the American pioneer, the beauty and utility of the forest was of little consequence. The aesthetics of the scene were subordinate to practical problems of clearing: simply, trees and stumps meant toil; cleared land meant production, food and neighbours; and work now meant the sacrifice of current well-being for a more glorious future. 'They have an unconquerable aversion to trees', said Isaac Weld:

> and whenever a settlement is made they cut away all before them without mercy; not one is spared; all share the same fate and are involved in the same general havoc ... The man that can cut down the largest number, and have fields about his house most clear of them, is looked upon as the most industrious citizen, and the one that is making the greatest improvements in the country.[13]

The United States was the classic example of land transformation through forest clearing. It is probable that over 46 million hectares had been cleared before 1850, the overwhelming bulk of which was carved out of the forests of the eastern half of the country, with only a very small amount coming from either natural or Indian clearings. It represents the culmination of two centuries of pioneering endeavour in the forests. Such clearing was a combination of 'sweat, skill and strength', and the pioneer farmer was seen as the heroic subduer of a sullen and unyielding wilderness which needed to be tamed.[14] The transformation of the forests was a universal and integral part of rural life. 'Such are the means', marvelled the French traveller the Marquis de Chastellux in 1789: 'by which North-America, which one hundred years ago was nothing but a vast forest, is peopled with three million of inhabitants ... Four years ago, one might have travelled ten miles in the woods I traversed, without seeing a single habitation.'[15] After 1850, the returns for 'improved land' in the federal censuses are a good indicator of the amount of land cleared. There was a big upswing in clearing in the ten years between 1850 and 1859, when a remarkable 16.1 million hectares was affected, equivalent to roughly one third of all the clearing that had gone on before, most of it in the north-eastern and northern Midwest states of New York, Ohio, Indiana, Illinois and Wisconsin, where 'every field was won by axe and fire'.[16] During the turbulent decade of the Civil War, the amount of forest cleared and settled dropped to 7.9 million hectares, but it rose again to its highest intercensal amount in 1870–9, when a colossal 20 million hectares was affected. After that, more acres of open prairie land were settled than of forested land, especially after the invention of the improved steel-tipped plough in 1837, which reduced the time taken to prepare ground for cultivation. Less than four days' labour were now needed to break the sod and plough one hectare of prairie, compared to nearly eighty days to clear a hectare of forest. Henceforth, forest clearing as an element in the formation of the landscape diminished in importance compared with other processes, both actually and in popular imagin-ation, as the glamour of the open range, cowboys – in fact, everything that made up the amorphous concept of 'the West' – overshadowed the often harsh reality of forest pioneering. The imperial conquest of the forest was replaced by the imperial conquest of the grasslands. In all, about 121.4 million hectares of forest were eliminated by the turn of the century, after which time forest reversion became a more dominant process than forest removal. It was one of the greatest deforestation episodes that the world has ever seen.

In addition, the industrial and domestic fuel demands on the forest were immense. The transformation of nearly 121 million hectares of forest into improved land after 1750 was accompanied by the vast amount of land cleared as a result of industrial logging from the second half of the nineteenth century onwards. Admittedly, a great deal of the wood was burnt on the spot; some was used for fences, houses and the like; some went into the manufacture of pearl ash and potash, but most of the remainder must have been burnt for heating the home. Simply, fuel wood was an indispensable requirement for existence in the cold, continental winters of the bulk of the United States.[17] Wood dominated as a source of energy, domestically and industrially, until well after 1885, and a case could be made for asserting that America's industrial growth in the latter part of the nineteenth century was based on cheap and abundant supplies of timber for fuel. Without the intensive use of its extensive forest, America could not have become such a major world power.[18]

Australia and New Zealand: settler empires

The United States was not the only place where European pioneers were hacking out a life for themselves and their families in the forest; it was also happening in the neo-Europes of Australia, Canada, and New Zealand. Australia's recent settlement, bureaucratic efficiency and seeming simplicity were not enough to give it the unequivocal documentation of deforestation that Marsh had hoped would be possible. Australia was no different from the other new countries, where 'the pioneers, whether settlers or timber-getters cut down indiscrimi-nately, giving no thought to anything but their immediate requirements'. The densest eucalypt forests are located in the humid areas in the extreme southern and eastern portions of the continent; as these were also the areas suitable for agricultural settlement, there was mindless destruction of much of the forest. The 'war of destruction', commented one observer, had reduced the forests of

New South Wales alone from 10.1 million hectares to 4.5 million hectares by 1900.[19]

Australia's eucalypts were especially difficult to clear because of the pre-dominance of hardwoods, the persistence of suckers and their ability to regenerate even after severe fire. Ringbarking, followed by the firing of the deadened trees, was the most common method of clearing.[20] The multi-stemmed mallee scrubland of South Australia, north-west Victoria, south-west Western Australia, south-western New South Wales and inland areas of Western Australia demanded a different method of clearing. Even if chopped down, the massive lignotuber root just below the surface remained alive and ready to sprout the next spring. The twin inventions of a primitive roller (1866) to knock down the slender scrub stems prior to burning and sowing in the ashes, and the stump-jump plough (c.1876), whose hinged and weighted shares kicked up harmlessly over the rocks and mallee roots, were the key to clearing millions of hectares of the lightly wooded areas of the continent, from the mallee lands of the south-eastern states to the brigalow scrubs of southern Queensland. Horse-drawn rollers could clear up to sixteen hectares a day; by the next century, that rate was achieved every hour when a heavy chain was dragged between a couple of crawler tractors.[21]

One can get no overall view of clearing in Australia; there is no comprehensive statistical series like the American census of 'improved' land. But the biographies of individual forests give a clue to what was going on. To take some examples: astride the New South Wales/Queensland border lay the 'Big Scrub', a rainforest of well over 4 million hectares which was described as 'all but impenetrable jungle, a lush profusion of vegetable growth'. Between 1880 and 1910, however, over 2.8 million hectares were cleared for dairying, sugar growing and timber. Equally large areas in the Strzelecki and Otway Ranges in Victoria were also cleared for dairying. One contemporary witness complained: 'The destruction of timber is enormous.' Everything was burnt on the spot and nothing was disposed of profitably, as transport links to markets had not been developed. The equivalent of the American wood-lot was never seen. 'Here the axe is set to every tree, and often not a shrub is left for shelter or cover.'[22]

Increasingly, technological advances had repercussions on land cover. The three inventions of the refrigerator (1880), the cream separator (1883), and the Babcock milk-testing machine (1892) led to perfection and expansion of but-ter-making, which, in turn, led to an expansion of dairying and the destruction of the forest in both Victoria and New South Wales/Queensland. It was even more evident with the development of dairying and pastoralism in remote New Zealand. Between 1840 and 1900, the forest cover was reduced from just over a half to about a quarter, most of that occurring between 1890 and 1900, when 3.6 million hectares, or approximately 14 per cent of the entire country, was cleared in order to establish a pastoral economy based on good-quality mutton, butter and cheese.[23] The previously perishable goods of one part of the empire were now transported with ease to the other parts, via the railways and refrigerated holds of the fast steamships. The 'tentacles' of empire were penetrating and linking every part of the world. There was a direct connection between the butter that British families spread on their toast for breakfast and the destruction of the great trees of the forests of Australia and New Zealand.

Agricultural and pastoral clearing were not the only inroads into the forest. Providing the wood for fencing, railways, mining, smelting and fuel in fledgling pioneer settler societies of only a few million consumed vast amounts of the best timber.[24] The upshot was that, by the end of the century, there was not enough timber left for housing and general construction in Victoria, especially after the excesses that followed the gold rush of the 1850s. In later years the cheap and abundant Oregon pine filled the shortfall for construction, with the termite-resistant redwood being used for railway sleepers. The same was true in the other countries around the Pacific Basin, such as China; even Peru and Ecuador proved to be lucrative markets for the enterprising lumber entrepreneurs of the Pacific Coast. Perhaps only New Zealand remained an exception, with its own flourishing timber industry. Now the Pacific Basin, like the Atlantic Basin, was fully enmeshed in the global timber trade.[25]

India: nationalistic empire

The human impact on the forests of India went back a long way. Long before the trading designs of the British East India Company culminated in 1857 in the establishment of the Raj – the pinnacle and epitome of imperial pomp, splendour, and territorial control – traditional societies had been vigorously exploiting their forests in a manner no more egalitarian or caring than that of Europe or Europeans overseas. For example, in the areas of Muslim overlordship in the north and west, pasturing and forest exploitation had been severe. In south-west India, shifting cultivation existed side by side with permanent agricultural settlement and village councils which regulated the amount of forest to be exploited by peasant farmers. Large landowners frequently dominated the forest use in their local areas; the forest was not regarded as a community resource. Scarce commodities such as sandalwood, ebony, cinnamon, and pepper were exploited under state and/or royal monopolies.[26] The signs of use, overuse, and ecological deterioration were to be seen everywhere.[27]

Over this traditional, but quite exploitative, society was placed a new imperial organization, from which we can isolate a number of innovations of deep significance for the ecology of the forests and their use. Three such imperial innovations are considered here.

Land survey, tenure and taxation

Land survey, the establishment of tenure, and the raising of taxes were interrelated processes in the imperial enterprise. Taxes were needed to support the colonial administration and to carry out the programme of public works which resulted from the 'modernization' process and concern about public health and food supplies. The taxes could not be levied, however, if land was not owned, the holding boundaries surveyed, and the owners 'fixed' in space. Shifting, fire-using *Kumri* agriculturalists seemed, to British eyes, to destroy wantonly prime timber like teak, and their manner of cultivation precluded taxation.

The unforeseen complication was that the imposition of high revenue assessments encouraged the expansion of internationally valuable cash crops, such as cotton, sugar cane and indigo, often at the expense of subsistence food crops. For example, over half of the 25,000 square kilometres of the south-eastern half of the doab below Delhi was covered by thick dhak (*Butea frondosa*) forests, often in continuous belts thirty kilometres wide. After the imposition of direct imperial rule and taxation in 1805, however, cultivation expanded into the forested interfluves and a vicious circle arose of raised rents, deeper debts, and more cultivation and forest destruction, all aided by 'natural' demographic pressures and increased economic activity. With the destruction of the forest, salinization became common, water-tables rose, many wells were abandoned, drainage channels dried out or were silted by moving soil, and malaria spread. The drought of 1837–8 dealt the final blow to the region.[28] It is easy to overstate the direct causal connection between these fiscal and economic factors and ecological deterioration, and it is also difficult to disentangle the consequences of population increase from the equation; nevertheless, imperial rationalization and deforestation did seem to go hand in hand.

Railways

The construction of railways was an integral part of British colonial military and commercial policy.[29] They were designed to strengthen the control of British rule with the speedy movement of troops and to augment British wealth with the export of raw materials (especially cotton) from India and the import of British manufactured goods into the subcontinent. By 1920, India's 99,130 kilometres of track were exceeded only by that of the USA, Russia and Canada. The internal development of the country *per se* was not the aim, but many railway promoters were conscious of the multiple 'modernizing' effects of railways, which they thought could not but be beneficial, especially in times of famine, when deficient areas could be supplied by surplus areas.

It is generally thought that the provision of transport, access to global markets, and halving of costs must have caused shifts in cropping patterns. But the evidence is ambiguous and inconclusive. Were the changes beneficial – by providing rising incomes – or were they deleterious – by encouraging the cultivation and export of non-food crops (like cotton) and even wheat, thus reducing the supply of domestic food? It seems that there was not a marked rise in incomes, nor a radical shift in cropping patterns.[30] Evidence from five large sample areas scattered across the subcontinent suggests that cotton showed a slight upward trend, from 6.4 to 9 per cent of cultivated land, and that food grains showed no major drop (they increased in many cases). It is even difficult to prove that food grain production *per capita* fell below consumption levels, except in the more extreme famine years.[31] Undoubtedly, the high price paid for cotton, which continued after the American Civil War, fed into the rural system and encouraged higher consumption and food production. The peasants did not *become* commercialized, many were *already* commercialized in the pre-colonial market economies of the Hindu and Muslim regimes. So that, by the time of the advent of the railway, the peasant proprietors were not the 'passive pawns of market forces', but canny opportunists who saw the chances for profit that the railways provided and were drawn into the global commercial market. Thus, the railways did not have the catastrophic effect on food supply that taxes did.[32]

Cultivation increased and the forest fell before the axe, particularly where there was direct European intervention. Clearing was extensive, for example, in the coffee plantations of the Wynaad plateau of Malabar, the tea plantations of the Assam hills, and the cotton-growing areas of Khandesh in the northern Deccan, though there was hardly any at all in the Deccan proper and Karnatak, where it might be most expected.[33] In the Irrawaddy delta region of Lower Burma, a conscious colonial policy of clearing the *kanazo* rainforests, draining the land and encouraging immigration was pursued by the British in order to grow rice to produce revenue and alleviate the periodic famines in neighbouring India. Between 1850 and the early 1900s, the population of the delta rose from 1.5 to over 4 million and it became one of the main rice-exporting regions of the world.[34] The result was the destruction of over 5 million hectares of rainforest.

In the immediate vicinity of the railway lines, the impact on the forest and the scrub jungles was severe. In Madras in 1856 the Conservator of Forests reported that:

> The wooded country occupying the notch between the Koondah and Anai-malai ranges was famous for wild elephants, but the extended cultivation along the line, and the increasing demand for wood, have jointly contributed to clear the primeval forests, and there is now only a thin scattered jungle.[35]

It was the same everywhere the railway lines went.

Shortages and forest policy

By the early 1860s, everything was moving inexorably towards a state of alarm about the reduction in the area and quality of the forest. An increase of over 38 million in the human population between 1880 and 1920, a bovine population that also increased by 35.6 million, and a concomitant demand for food and fodder and all-too-frequent famines meant that the quest for more land could never be relaxed. In the absence of any major trends towards intensification, expansion into the forest was a deliberate part of British land policy, especially as it helped to maximize revenue.

But there was a problem. In addition to providing 'new' land, the forest performed many important 'free' functions for the existing rural villager and shifting cultivator alike, supplying green forage, especially in times of drought, wood for construction and fuel, wild fruits, nuts, honey and game, and the land in which to shift the cultivation cycle. On top of this, the forest was the source of essential fuel and constructional supplies for the railways, widening the radius of fuel provision far beyond what had been reasonable or profitable before.[36]

In response to such pressures, the British authorities embarked upon one of their great 'modernizing' projects: to suppress the 'vagabond' agricultural and fire habits of the itinerant *Kumri* cultivators in the hills and to set aside forest reserves of valuable commercial timbers for preservation, renewal and management by professional foresters. Even the agriculturalists were not immune to the new purist regime. Like suttee or smallpox, the indiscriminate cutting of commercial trees for fuel, constructional timber and cultivation was 'an evil that must be suppressed'.[37] It was an intricate, circular, self-reinforcing and self-justifying argument that the stewardship of the forests and the control of the hill people were essential to the survival and welfare of the mass of Indians.

The result was the massive administrative and management structure of the Forest Department, which acquired one fifth of the total land of the subcontinent by the end of the century.[38] Forests were protected to secure timber, and it was now possible to exclude rural land users. But one is left with a terrible paradox. An administrative edifice had been constructed which had few parallels in the world at this time or for decades to come. It may have been one of the administrative jewels in the imperial Crown, a model for the rest of the world and a highly efficient and profitable enterprise, but it is also one of the festering sores in the body of the subcontinent, one that has yet to be healed.[39] The foresters and their regulations became the face of alien power which pervaded Indian rural life just as surely as any military, judicial or political administrative framework, and they cut deeply into the fabric of traditional life.

China: autonomous continental empire

The frontiers of the neo-European lands were paralleled by the vast inland frontier of China, a country of semi-continental proportions, almost untouched by western influences: an autonomous, imperial world unto itself. Unlike India, little information on deforestation exists; it was, said one writer on forest depletion, a land of 'ponderous unknowns'.[40] Two things are certain, however: the population was large (65–80 million around 1400, and 385 million in 1893); and deforestation was of great antiquity and extent.[41] Given our knowledge of comparable societies at comparable times, it seems inconceivable that an increment of 300 million people, and the 'new' agricultural land that would have been necessary to feed them did not have a dramatic effect on the natural vegetation, particularly the forest. Although some raising of yields can be detected, it is more likely that cropping was increased first by taking in sub-marginal land on the forested slopes of south and central China, dryland on the northern fringes and Inner Mongolia, and wetland in the river bottoms and deltas.[42] What we do know is that, between 1700 and 1920, the increment of 66 million hectares of new cropland led to the destruction of over 56 million hectares of forest land (see Table 11.1).

The Qing Dynasty (1644–1911) was the archetypal imperial regime, highly centralized, bureaucratic, and expansionist. An all-powerful mandarin bureaucracy dominated the multitude of autonomous village communities. Dominance over society was ensured through the management of public works and an active role in migration, land clearance and reclamation, sometimes through corvée labour. Although remaining obstinately feudal, the state did not attempt to interfere in the daily affairs of the villagers. At a local level, it tolerated a fairly liberal and flexible market system which encouraged a merchant class, credit, and thriving market towns, but it also directed major land reclamation projects of irrigation, drainage, and clearing as part of a dynastic concern for the welfare of its subjects and the feeding of an expanding population.[43] The large (210,000 square kilometres) province of Hunan in southern central China is a good example.[44] During the later Ming Dynasty (1386–1644) and the early part of the Qing Dynasty state-sponsored land clearance and immigration of Han peoples from the north were initiated in order to increase the tax base, enhance local state revenues, and support a bigger bureaucracy and militia. In the process, Hunan changed from an underpopulated peripheral province to a new settler society on the edge of the densely settled imperial core of the country. The population rose from about 5 million in 1582 to 14.9 million in 1776 and 20 million in 1842, while the area of cultivated land rose from 1.8 to 2.6 to over 4 million hectares, much of it coming out of the forests on the hill slopes.

By the late eighteenth century, it seemed as though expansion into the forested lands of Hunan had overstepped the mark everywhere. Cultivation was no longer confined to rice production in the flat river-valley bottoms; the introduction of New World crops such as potatoes and maize had allowed 'slash-and-burn' cultivation to move up steep slopes.[45] Hillsides were 'stripped of trees', there was a scarcity of firewood, and rivers were swollen with silt:

> When there are too many people there is not enough land to contain them, there is not enough grain to feed them, and there are not enough wealthy people to take care of them ... Water rushes down violently from the mountains and the mountains became bare; earth is dug out ... and the rocks are split; there is not an inch of cultivable land left, while lowlands were silted over.[46]

Throughout the eighteenth and nineteenth centuries China moved inexorably towards the almost total deforestation of its domain; by the beginning of this century, it was an impoverished peasant society, subsisting in a treeless, denuded and eroded lunar landscape. The country was held up to the world as the prime example of what excessive deforestation could do.[47]

Japan: autonomous island empire

The contrast with neighbouring Japan could not have been more stark. Everything we know about that country suggests the creation of an ordered and moderately affluent peasant society, subsisting on the multiple bounty of forest (which supplied timber, leaf fodder, fertilizer and fuel), land and sea.

Japan had gone through an intense and serious timber crisis between the fifteenth and mid-seventeenth centuries. Phases of intense predatory logging were authorized by the regional military and feudal lords (*daimyōs*) in order to construct towns, monuments and fortifications as they feuded constantly amongst themselves. Unauthorized logging also damaged the forests severely. A demoralized peasantry cultivated excessively on steep slopes, disputes over forest use were common, and erosion, flooding and sedimentation of the cultivable lowlands caused alarm. The quantity and quality of timber deteriorated, and scarcity of supply became widespread. It was remarkable that these multiple problems were recognized as early as the mid-seventeenth century by the Tokugawa shogunate or bakufu, which gradually imposed a tight control and hegemonic rule over the *daimyōs*, who eventually became vassal lords. It is even more remarkable that, through coercion and a recognition of self-interest, all took positive measures to control and alter forest use.[48]

It is difficult to generalize about the remedial measures when one realizes that the Tokugawa *bakufu* lasted for over 250 years and that there were 250 or so *daimyōs*, but one can say that the imperial administration devised what Totman called a 'negative regimen'.[49] Administrative rules were evolved for both the protection of the forests and the safeguarding of products. The rights of villagers and lords and the forest domains in which exploitation could take place were carefully delineated and codified; the emphasis was on 'rights', as in Tokugawa Japan the concept of land 'ownership' did not exist. Regulations were devised to control the areas open to harvest, the number of days and the number of workers who had access to the area, the size and type of tools to be used, the number of loads that could be produced, the inspection of the transport of timber goods, and the size of buildings and type of timber to be used. Sumptuary regulations prohibited the use of certain prized woods like sugi (*Cryptomerica japonica*) and hinoki (*Chamaecyparis obtusa*) or Japanese cypress. A comprehensive, if uncoordinated, body of regulations constituted 'a vast and moderately effective system of rationing'. Protection was extended to lowland basins, and there was even the creation of an Office of Erosion Control to supervise excessive cutting in the Kinai Basin. In many ways, the 'negative regimen' solved nothing during this era of growing population, but it was an essential prelude to raising awareness and knowledge, 'buying time' for the introduction of more effective policies of silviculture and the purposeful planting of trees to produce the 'green archipelago' of contemporary Japan, rather than the ruined land that might have been expected.

In an attempt to understand this remarkable precocity in forest matters,

Totman dismisses such facile explanations as a 'love of nature', or Buddhist, Confucian or Shinto sentiment for natural objects; Japanese sensibility to nature was always refined, delicate, recreational and purely urban in orientation, and trees were crude and rural. In any case, these religious values had not saved the forests of the pre-Edo era (nor had they saved them in mandarin, Confucian China).[50] Rather, the concerns of those who wanted to restore the forests were intensely practical; forests helped to supply human needs for raw material, fertilizer and land, aims which were aided intentionally and unintentionally by the natural succession of deciduous trees where coniferous trees had flourished before. It was a form of conservation ethic. Additionally, for some unknown reason, wheeled vehicles and cross-cut saws were rarely permitted in forested areas, a constraining of technology which probably helped preservation, as did the absence of sheep and goats. Finally, Totman suggests that the overriding concerns of the Tokugawa administration to preserve peace and avoid foreign contacts focused attention on the resources in hand and prevented the introduction of alien, destabilizing technology and values. There is a significant lesson in this. It was a unique social and environmental situation in which a disciplined and literate society sorted out its priorities. The people of Tokugawa Japan, high and low alike, had to make do with what they had and what they could acquire peacefully, and they knew it.[51] The fact that population stabilized after 1750 certainly took pressure off the forests; the search for protein and resources was extended out into the ocean, not into the forest. As a result, the total demand for woodland actually decreased. It was a remarkable experiment and example, with few, if any, counterparts in any part of the world.

Conclusion

This necessarily cursory view of the relations between imperialism and ecology suggests that it was a far more complex and varied phenomenon than much current writing would have us believe. Imperialism must not be equated narrowly with western colonialism; it is a more useful and generic geographical concept that includes exploration, organization and settlement as well as exploitation. The legacies of imperialism are many: there is political absorption, economic change and redirection, racial/ethnic mixing and stratification, acculturation, and social and psychological domination – to mention the most obvious – all of which have created 'peoples without history'.[52] But the bequest of major ecological change, a new sort of 'ecological imperialism', is all too readily overlooked, and deforestation is one of its most blatant manifestations. With the exception of Japan, the felling of the forest of the countries of the Pacific Rim continued throughout the nineteenth century and reached a new crescendo after 1950.

Notes

1. Alfred W. Crosby, *Ecological Imperialism* (Cambridge: 1986).
2. See Alfred W. Crosby, 'Ecological Imperialism: The Overseas Migration of Western Europeans as Biological Phenomenon', in Donald Worster (ed.), *The Ends of the Earth* (Cambridge: 1988), pp. 103–17. Attention should be drawn to the pioneering work of the Australian writer Sir A. Grenfell Price, whose *The Western Invasions of the Pacific and its Continents* (Oxford: 1963), and *The Importance of Disease in History* (Adelaide: 1964), were early, but neglected, manifestations of this genre.
3. See Carl O. Sauer's early and uncompromising critique of ecological and cultural destruction in his 'Theme of Plant and Animal Destruction in Economic History', *Journal of Farm Economics*, 20 (1938), pp. 765–75, and 'Destructive Exploitation in Modern Colonial Expansion', *Comptes Rendus du Congrès International de Géographie,* vol. 2 (Sect. 3c) (Amsterdam: 1938), pp. 494–9.
4. Daniel R. Headrick, *The Tools of Empire* (Oxford and New York: 1981), and *The Tentacles of Progress* (Oxford and New York: 1988).
5. Claims to novelty should be tempered with the knowledge that, in 1966, D. K. Fieldhouse wrote *The Colonial Empires* (London), which the author claimed (probably correctly) provided 'a prospectus for a virtually new field – the comparative study of empires' (p. xiv), but the environment was certainly not a part of that prospectus.
6. George Perkins Marsh, *Man and Nature: or, Physical Geography as Modified by Human Action* (New York: 1864), p. 186.
7. Marsh, *Man and Nature,* p. 49.
8. For the distribution of forests and their characteristics, see Geoffrey W. Leeper (ed.), *The Australian Environment* (Canberra: 1970), pp. 44–67, 120–30; D. A. N. Cromer, 'Australia', in S. Haden-Guest, J. K. Wright, and E. M. Teclaff (eds), *A World Geography of Forest Resources* (New York: 1956), pp. 573–90.
9. See Michael Williams, 'Thinking about the Forest: A Comparative View from Three Continents', in Susan L. Flader (ed.), *The Great Lakes Forest: An Environmental and Social History* (Minneapolis: 1983), pp. 253–73; John Dargavel and Richard Tucker (eds), *Changing Pacific Forests: Historical Perspectives on the Forest Economy of the Pacific Basin* (Durham, NC: 1992).
10. Donald W. Meinig, 'A Macrogeography of Western Imperialism: Some Morphologies of Moving Frontiers of Political Control', in Fay G. Gale and Graham H. Lawton (eds), *Settlement and Encounter: Geographical Studies Presented to Sir Grenfell Price* (Melbourne: 1969), pp. 213–40.
11. Table 11.1 is based on John F. Richards, 'Land Transformation', in B. L. Turner, II, *et al.* (eds), *The Earth as Transformed by Human Action* (New York: 1990), p. 164.
12. Michael Williams, *The Americans and their Forests: A Historical Geography* (New York: 1989), pp. 357–61.
13. Isaac Weld, *Travels Through the States of North America and Provinces of Upper and Lower Canada During the Years 1795, 1796 and 1799,* 2nd edn (London: 1799), vol. 1, pp. 231–3.
14. David M. Ellis, *Landlords and Farmers in the Hudson–Mohawk Region, 1790–1850* (Ithaca, NY: 1946), p. 73.
15. François J. Chastellux, *Travels in North America in the Years 1780, 1781 and 1782* (New York: 1789), p. 47.
16. Daniel Millikin, 'The Best Means of Preserving and Restoring the Forests of Ohio', *Ohio Agricultural Report,* 2nd Ser. (1871), p. 319.
17. R. V. Reynolds and A. H. Pierson, *Fuelwood Use in the United States, 1630–1930,* USDA Circular No. 641 (Washington, DC: 1942). For other calculations, see S. H. Schurr and B. C. Netschert, *Energy in the American Economy: An Economic Study of its History and Prospects* (Baltimore: 1960), p. 5.
18. Williams, *Americans and their Forests,* pp. 146–57, 331–44.

19. T. A. Coghlan, *The Timber Resources of New South Wales* (Sydney: 1900), p. 1; R. Kaleski, 'Our Forests in Earlier Days: Some Political History', *Australian Forestry Journal*, 8 (1925), pp. 323–6.

20. R. T. Archer and P. J. Carrol, 'Dairy Farming', in *Victoria: The Yearbook of Agriculture* (Melbourne: 1905), pp. 297–333; S. M. Wadham and G. L. Wood, *Land Utilization in Australia* (Melbourne: 1939), pp. 61–73, 111–12.

21. For a thorough discussion of mallee clearing, see Michael Williams, *The Making of the South Australian Landscape* (London: 1974), pp. 124–77.

22. T. A. Coghlan, *Picturesque New South Wales: An Illustrated Guide for the Settler and Tourist* (Sydney: 1901), p. 90; A. M. Laughton and T. S. Hall, *Handbook to Victoria* (Melbourne: 1914), pp. 41, 331.

23. Kenneth B. Cumberland, 'A Century's Change: Natural to Cultural Vegetation in New Zealand', *Geographical Review*, 31 (1941), pp. 529–54; Andrew H. Clark, *The Invasion of New Zealand by People, Plants and Animals* (New Brunswick: 1949).

24. See Michael Williams, 'The Clearing of the Woods', in R. L. Heathcote (ed.), *The Australian Experience: Essays in Australian Land Settlement and Resource Management* (Melbourne: 1988), pp. 115–26.

25. See Thomas R. Cox, *Mills and Markets: A History of the Pacific Coast Lumber Industry to 1900* (Seattle: 1974), esp. pp. 214–22.

26. M. V. Nadkarni, with S. A. Pasha and L. S. Prabhakar, *The Political Economy of Forest Use and Management* (New Delhi: 1989), pp. 29–36.

27. Madhav Gadgil and Ramachandra Guha, 'State Forestry and Social Conflict in British India', *Past and Present*, 123 (1989), pp. 141–77.

28. Michael Mann, 'Ecological Change in North India: Deforestation and Agrarian Distress in the Ganga-Jamna Doab 1800–1850', *Environment and History*, 1/2 (1995), pp. 201–20.

29. The most accessible general accounts of the railways are in John M. Hurd, 'Railways', in Dharma Kumar (ed.), *Cambridge Economic History of India*, vol. 2: *c. 1752–c. 1970* (Cambridge: 1983), pp. 737–61; Headrick, *The Tentacles of Progress*, pp. 49–95.

30. Michelle B. McAlpin, 'Railroads, Prices, and Peasant Rationality: India, 1860–1900', *Journal of Economic History*, 34 (1974), pp. 662–3.

31. Michelle B. McAlpin, 'Railroads, Cultivation Patterns, and Foodgrain Availability: India, 1860–90', *Indian Economic and Social History Review*, 12 (1975), pp. 46–7, 52–3.

32. McAlpin, 'Railroads, Prices, and Peasant Rationality', p. 683; John F. Richards and Michelle B. McAlpin, 'Cotton Cultivation and Land Clearing in the Bombay Deccan and Karnatak, 1818–1920', in Richard P. Tucker and John F. Richards (eds), *Global Deforestation in the Nineteenth-century World Economy* (Durham, NC: 1988), p. 80.

33. Based on Richard P. Tucker, 'The Depletion of India's Forests under British Imperialism: Planters, Foresters and Peasants in Assam and Kerela', in Worster (ed.), *The Ends of the Earth*, pp. 118–41; Richards and McAlpin, 'Cotton Cultivation and Land Clearing', pp. 75–93.

34. Michael Adas, 'Colonization, Commercial Agriculture and the Destruction of the Deltaic Rainforests of British Burma in the Late Nineteenth Century', in Tucker and Richards (eds), *Global Deforestation*, pp. 95–111.

35. Quoted in E. P. Stebbing, *The Forests of India*, vol. 2 (London: 1922), p. 99.

36. George Bidie, 'Effects of Forest Destruction in Coorg', *Journal of Royal Geographical Society of London*, 39 (1869), p. 88.

37. B. H. Baden-Powell, 'The Political Value of Forest Conservancy', *Indian Forester*, 2 (1876), p. 280.

38. Bertold Ribbentrop, *Forestry in British India* (Calcutta: 1900), p. 72; see also pp. 76ff., for the subsequent organization of the department.

39. K. Sivaramakrishnan, 'Colonialism and Forestry in India: Imagining the Past and Present Politics', *Comparative Studies in History and Society*, 37 (1995), pp. 3–40.

40. Rhoads Murphey, 'Deforestation in Modern China', in Tucker and Richards (eds), *Global Deforestation*, pp. 111–28.

41. D. Perkins, *Agricultural Development in China, 1386–1968* (Chicago: 1969), table 2.1 and pp. 15–16, Appendices A and B, pp. 192–240. For earlier studies, see Ping-ti Ho, *Studies in the Population of China, 1386–1953* (Cambridge, MA: 1959), esp. pp. 1–97.

42. Perkins, *Agricultural Development in China*, pp. 13, 38–53; Murphey, 'Deforestation in Modern China,' n. 2.

43. Francesca Bray, 'Agriculture', in J. Needham (ed.), *Science and Civilization in China*, vol. 6: *Biology and Biological Technology, II: Agriculture* (Cambridge: 1984), pp. 94–6.

44. See Peter C. Perdue, *Exhausting the Earth: State and Peasant in Hunan, 1500–1850* (Cambridge, MA: 1987), pp. 41–58, 64–80, and tables 3 and 4.

45. Other 'hill' migrations in southern China that produced extensive deforestation are outlined in E. S. Rawski, 'Agricultural Development in the Han River Highlands', *Late Imperial China* (formerly *Ch'ing-shih wen-t'i*), 3/4 (1974), pp. 63–81.

46. Perdue, *Exhausting the Earth*, pp. 35–6, 88.

47. William C. Lowdermilk, 'Forestry in Denuded China', in H. F. James (ed.), *China, Annals of the American Academy of Political and Social Sciences*, 152 (1930), p. 127.

48. Based on Conrad Totman, *The Green Archipelago: Forestry in Preindustrial Japan* (Berkeley: 1989), pp. 34-80.

49. See Totman, *Green Archipelago*, pp. 83–115.

50. See Yi-fu Tuan, 'Discrepancies between Environmental Attitude and Behaviour: Examples from Europe and China', *Canadian Geographer*, 12 (1968), pp. 176–91.

51. Totman, *Green Archipelago*, pp. 184, 179–80.

52. Eric Wolf, *Europe and the People without History* (Berkeley: 1982).

CHAPTER 12

Global developments and Latin American environments

Elinor G. K. Melville

In a recent article, Robert Marks writes that 'the story of environmental change in [south China] raises the question of whether a driving force of modern global environmental change is the "rise of the capitalist mode of production", or a broader, more general process of commercialization, regardless of what mode of production sustains it.' He calls for environmental historians 'to develop a global perspective on environmental change' in order to clarify the relationship of global processes to local environmental processes over the past 500 years.[1] This chapter is a preliminary contribution to this project.

As Marks notes, the concept of the 'rise of the capitalist mode of production' is used by environmental historians to explain environmental change in many parts of the world over the past 500 years. In this model of global development, capitalism is equated with the international market, and environmental change after 1500 is the consequence of the spread of European commercial activities.[2] A variant of this model equates the growth of the international market with the development of a world system structured by a global division of labour and unequal exchange between centre and periphery.[3] In both these models, the contemporary twentieth-century world order is seen to be the inevitable consequence of European expansion; and environmental change in both the formative era and the era of consolidation of this world order (the sixteenth and the nineteenth centuries respectively) is perceived as essentially the same – that is, over-exploitation and destruction of natural resources in an attempt to meet Europe's needs.

This chapter examines whether the 'rise of the capitalist mode of production' can be used to explain environmental change in continental Latin America over the conquest and colonial eras, from 1519 to about 1810, roughly the same time-span that Marks examined. I argue, like Marks, that the changes observed in the Latin American environments over this period were not associated with the rise of the capitalist mode of production, if by this is meant the development of a global system of exchange and production *that is structured by a global division of labour*. I will use recent historiography of Latin America to demonstrate that, while there was, as in China, increased commercialization during this period, the global division of labour did not appear until the nineteenth century and, therefore that 'the rise of the capitalist mode of production' cannot be used to explain environmental change outside Europe prior to this century. My primary focus is thus on economic change, but I believe that, if we are going to use economic models to explain environmental change, it is necessary to clarify these models and our use of them.

Two notes of explanation before I start. Contrary to the rationale for this book, I will be dealing explicitly with the indigenous peoples. This is because, in contrast to the so-called 'settler societies' or, better, the 'lands of demographic takeover', indigenous peoples and indigenous plants continue to be major elements of Latin American landscapes and environments; and because the historiography of early Latin America has concentrated principally on the problems of conquest and domination of huge populations by a very small group of invaders and on the actions of the indigenous peoples in the formation of the colonial regimes. The discussions of the conquest and colonial eras thus focus primarily on the centres of high population density: the *Audiencia* of New Spain that was founded on the former Aztec Empire; and the *Audiencia* of Peru founded on the Inca Empire. These regions formed the centres of European power in Latin America until the nineteenth century, and they distinguish Latin America from other examples of early modern European imperialism. When we consider the nineteenth-century global processes, the focus moves to the fringes and frontiers of Latin America, thereby mirroring a historical shift of the economic centre of gravity of the region from the old colonial centres in the highlands of Mexico and Peru to the developing livestock frontiers in the sparsely populated Atlantic coastal regions.

The second note has to do with sources, but it will also serve to move us into the subject of the chapter. Self-conscious Latin American environmental history is relatively new, and this chapter is based on secondary sources written by historians who were not thinking of human–nature relations in terms of a dialogue – or of the reciprocal influence of natural and social process. As a result, we can pick up both some of the perceptions that have informed historians' thinking about nature and also, perhaps, the thinking of those who have used historical writings as resources – and hence the underpinnings of the idea that environmental change in the formative sixteenth century was, as in the nineteenth century, a process of environmental destruction.

Perceptions of human–nature relations in Latin American history

There are some intriguing shifts in historians' perception of nature in the regions that became Latin America. In the histories of the conquest era (1492/1519–c.1580), for example, 'nature' is seen as the victim of human/European aggression, and, because the indigenous peoples were the most directly affected by the introduction of Old World species – above all, by the new disease organisms – they are also seen as victims and, by extension, as part of 'nature'. In the histories of the colonial era (1580–1810), by contrast, 'nature' is dangerous and capricious: the once aggressive Europeans are now perceived as part of a settled colonial world, where they and the Indian peasant communities face a common foe: the unpredictable physical environment. These studies take agrarian histories of Europe as their models, and use concepts such as natural hazards, subsistence crises, and the moral economy to discuss the interface between humans and their physical environment.[4]

Traditionally, the colonial era has been characterized as a period of stagnation; despite evidence of social, cultural, and economic change, the perception of stability has persisted. This has had the effect of providing a context for the study of social reproduction, and environmental historians of the colonial era have recently begun to study the continuous adjustments that result in the maintenance of characteristic patterns of human–nature relations: that is, the reproduction of environments, rather than their transformation.[5] Latin American environmental history is thus developing two quite distinct approaches: students of the conquest take Merchant's concept of ecological revolutions as their model, and see human–nature relations in terms of rapid, revolutionary transformations; students of the colonial era take a more Braudelian approach, viewing human–nature relations in terms of cycles and reproduction.[6] In this sense, Latin American history provides a significant contrast to US environmental history, which reflects both the historical experience of a continuously moving frontier and the thesis that it spawned, tending to view environmental change as open-ended, transformative and declensionist.

When we move on to the nineteenth century, 'the independence era' as it is known in Latin American historiography, we find perceptions of human–nature relations shift yet again. As in the sixteenth century, the environments of Latin America are seen to be under enormous stress; once again, nature is perceived as the victim of aggressive human/European behaviour.[7] Despite the historiographical similarities, environmental change in the sixteenth and the nineteenth centuries could not be more different. Sixteenth-century environmental change was marked by the transformation of native ecosystems, the most striking – or simply the best-studied – result of this transformation being the appalling loss of human life through the introduction of alien disease organisms. By contrast, nineteenth-century change was marked by the over-exploitation of natural resources and a dramatic increase in human populations. Not only were the environmental changes in the sixteenth and the nineteenth centuries different in themselves, but they were associated with very different global processes: sixteenth-century changes were associated with the biological integration of the Americas with Eurasia and Africa; nineteenth-century changes were associated with the economic integration of the Latin American economies into a global division of labour as a resource supply region.[8]

The conquest and colonial eras, 1519–1810

Two distinct fields in the historiography of Mexico have had a remarkable impact on the way in which Latin American historians view the effect of the

European invasion: the first is epidemic and demographic history; the second, ethnohistory. Beginning in the late 1940s, Woodrow Borah and Sherburne Cook began a series of studies of the impact of the sixteenth-century pandemics on the indigenous populations of the region that became New Spain; they demonstrated a demographic collapse of the order of about 90 per cent over the first 100 years. As a corollary, they argued that the indigenous populations at contact were very large, thereby initiating a debate over the size of the pre-contact populations and the impact of the Europeans which continues, with unabated enthusiasm – and not a little fulmination – to this day.[9]

By demonstrating the close relations between economic and social change and changes in the disease environments of the region, they can be said to be the 'founding fathers' of Latin American environmental history. Their work pointed up what is perhaps the most striking difference between the Latin American and the North American colonies – and hence the difference in approaches of Latin American and US historians to the study of indigenous influence on the formation of the modern environments – the presence in 1519 in Central and South America of centralized states and empires encompassing millions of people. The empire of the Inca, for example, stretched along the backbone of the Andes from the equator to 75° south. The population of the Meso-American region, most of them sedentary farmers, is estimated to have been around 21.4 million people at contact, with a further 11.5 million in the Andean region.[10] The extraordinary density of these agricultural populations and, more importantly, their organization into social systems that funnelled surpluses out of the peasant producing communities to a centralized urban élite have, over the past half-century, come to give a very specific cast to the historiography of conquest and colonization in the Spanish American world.[11]

The second field, ethnohistory, began in earnest with the publication in 1967 of Charles Gibson's classic work, *The Aztecs under Spanish Rule*.[12] Prior to the publication of Gibson's work, the conquest was generally viewed as an event: the Spaniards came, saw and conquered, and then got down to the business of governing the indigenous peoples, extracting the wealth of their land and labour and funnelling it to Europe. Gibson's work, and the increasing numbers of ethnohistorical studies that it stimulated, changed our understanding of the conquest: rather than an event, the conquest is now treated as a process that took decades – in places, centuries – to be played out. It has become clear that the Spaniards could not do exactly as they wished; negotiation, as well as violence, was needed to win labour and land. Indeed, it is a truism in recent historiography of the conquest era that the Spaniards had to adapt to local realities, at least during the formative decades of the sixteenth century. A couple of examples will demonstrate this point.

In the early decades of the conquest of New Spain, when Spanish *conquistadores* had only indirect access to land,[13] Old World crops were grown on Indian lands using Indian labour, very often according to indigenous management systems and schedules: wheat, for example, was cultivated in the mounds that were typical of maize agriculture.[14] Despite the fact that wheat was an essential element in the culture of the *conquistadores* and was grown on Spanish-owned haciendas later in the sixteenth century, there were two reasons why the Spaniards could not simply impose the requirements of their land management systems: first, wheat production on a commercial scale competed with Indian work schedules, particularly maize cultivation; second, the best grains for bread could not be grown successfully during the summer rainy season because of rust infections. So the Spaniards changed their scheduling: from spring sowing of rain-fed wheat and an autumn harvest, to autumn sowing of irrigated wheat and a spring harvest. The shift to a spring harvest meant that threshing had to be postponed to the following autumn when the traditional method of threshing with large numbers of mares on a flat area of hardened ground was not threatened by early rains and did not use animals that were needed to prepare the ground for maize. This postponement of threshing, in turn, meant that the Spaniards had to build enormous stone barns to store the wheat both before and after threshing. Adaptation to local socio-economic systems and local climates meant, in this case, a considerable capital outlay.[15]

Further complicating our understanding of the consequences of the European invasion for Latin American environments was the selective adoption of European cultural elements by the indigenous peoples. The novelties introduced by the Europeans were not spurned by the indigenous peoples – why should they be, if they were useful? Indigenous peoples added animals, plants, and land management practices to their cultural array, apparently with few qualms. The adoption of Old World animal species is a case in point. Pastoralism was found only in the Andean highlands before the arrival of the Europeans in the Americas; only a few decades after initial contact, however, a multitude of new forms of agro-pastoralism had appeared on the American scene, giving the American landscapes a distinctive and characteristic 'look'. The formation of these new agro-pastoral systems was not unidirectional, however: Europeans were the agents of the initial introduction of the new animal species, but recent studies suggest that the strongest influence on the development of American agro-pastoralism was a combination of American ecosystems and indigenous societies and cultures. While the initial invasion influenced the selection of the species of animal introduced, and very often the specific breed, the process did not stop there: creole breeds developed *in situ* that were adapted to the local ecosystems and to the sociocultural and economic needs of the local human populations.[16]

Where an introduction was not perceived to be a suitable addition to, or replacement for, an indigenous cultural item, it was not adopted. The problems faced by the Spaniards in their attempts to force indigenous peasant communities to grow wheat for tribute is a good example of the ability of those communities to avoid growing something for which they had no use.[17] Indeed, wheat was never accepted as a basic staple by the indigenous peoples of Latin

America, and the predominance of indigenous grains and other staples is characteristic of modern 'indigenous' landscapes of the neotropics. The processes
by which the Spaniards moved into, and ultimately dominated, indigenous
peoples and places were not characterized by the straightforward imposition
of European economies, nor even of Old World land management systems –
despite the introduction of Old World animal and plant species. Equally, the
formation of Latin American landscapes and environments cannot be viewed
as a simple replacement of indigenous with European forms.

The colonial regimes that evolved over the sixteenth century in the central
highlands of Mexico and Peru were based on a seemingly inexhaustible supply
of indigenous labour and *increased* commercial activity. The Europeans did not
introduce commerce into these regions; rather, they increased the extent to
which market exchange shaped social and economic relations.[18] The Spaniards
found complex systems of exchange that included reciprocal exchange, centralized state redistributive systems, market systems, and long-distance trade, and
that linked regions of very different cultures, political economies, and ecosystems. Indeed, specialization of production and exchange between the
remarkably varied ecosystems of the Central and South American continent
was a hallmark of pre-Hispanic societies. Evidence from both archaeological
and documentary records indicates that long-distance trade extended over vast
stretches of sea and land at the time of the European invasion.

The clearest evidence of what is commonly thought of as 'commercial practice' in the pre-Hispanic era comes from Meso-America, the region that later
became Mexico, Guatemala, and Belize, and is found in the descriptions gathered by a Franciscan, Fr. Bernadino de Sahagún, of the huge daily market of
Tlatelolco in the Aztec capital (the Spaniards estimated a daily attendance of
50,000 in Tlatelolco in 1519, a figure which indicates relative size rather than
actual numbers). An extraordinary variety of subsistence, artesan, and long-
distance goods were presented for sale; prices were set daily, using gold as a
standard of value; media of exchange (most commonly, cacao beans and pieces
of cloth) with a set exchange value were used to acquire goods; and market
police controlled the quality of goods on sale.[19]

It is clear from their actions after the arrival of the Spanish that the inhabitants
of the centralized states, if not the more egalitarian societies, lacked neither a
desire for, nor a knowledge of, commerce, nor did they lack a drive for the
acquisition of wealth; when the Spaniards opened up new opportunities for
commerce, they took immediate advantage of them. Writing of the Andes in
the decades immediately following the arrival of the Spaniards, Steve Stern
notes that individuals 'reacted innovatively to the new colonial economy'; far
from 'participating reluctantly in the commercial economy just to gather
moneys needed for tribute', indigenous communities 'displayed an open,
aggressive – even enthusiastic – attitude' towards the opportunities that were
provided by Spanish commerce. He cites the case of a community that took
advantage of the labour market in the silver mines of Potosí, sending workers

100 kilometres to work there, before the Spaniards gained control of the community.[20]

The Spaniards found it far more profitable to cream off the profits generated by indigenous systems of production and trade; it was only with the collapse of the indigenous populations that we find them making a concerted effort to set up their own agricultural production systems. (Precious metals were their main interest; they used indigenous labour in the first decades to pan for gold, turning to silver mining when they located the huge silver mines in Mexico and Peru.) Local élites, at least at first, seem to have welcomed the Spaniards as sources of power which they could manipulate to their own ends, increasing their control over the subject populations and generating a larger surplus, part of which was paid to the Spanish officials and part of which was used in the new trading opportunities. The active participation of the indigenous élites not only enabled the generation of a surplus which supported the Spanish royal officials and provided those officials with the means to amass considerable wealth, but also enabled the Spanish government indirectly to control huge populations. By such means, the Spaniards built a commercial system that was, at the same time, a system of colonial government.[21]

Robert Patch suggests that the manner in which the Spanish officials inserted themselves into the indigenous systems of production, manufacture and distribution of cotton and textiles in Central America resembles the better-known 'country trade' of the Far East. Patch also notes that their activities were not intrinsically alien to the indigenous world, nor were they inherently destructive of that world:

> Economic exploitation and commercialization, while disruptive in some contexts, were perfectly compatible with Indian traditions, and thus native culture had a better chance of adapting to colonialism, rather than succumbing to the excessive demands of the colonists. To this day, the Guatemalan and Chiapan highlands – the textile manufacturing regions containing the most lucrative magistracies in the kingdom – are among the most Indian parts of Latin America.[22]

Systems of production, manufacture, and exchange that were built on an indigenous infrastructure were clearly responding to local needs. But the systems of production and manufacture developed by the Spaniards were also aimed at local regional and inter-regional markets, and early trade flows between Spain and its colonies were reduced as the colonists produced their own woollen cloth, furniture, ships and, in the sixteenth century, wine and silk. The decline in trade hurt the Spanish merchants far more than the colonial; the Crown responded by increasing monopolies, denying the colonies the right to produce such things as wine and olives, and prohibiting inter-regional trade, a move which resulted in a spectacular increase in contraband.[23]

The Latin American internal markets are now thought to be as important, if not more important, than silver production, as the motor driving the growth of the colonial economies. The silver flows themselves indicate that, by the beginning of the seventeenth century, the centre of gravity of the colonial economies was America, not Europe: by this date, increasing amounts of silver remained in America to be invested there; silver was (illegally) carried west in the Manila galleon to buy silks and porcelain for sale in the Americas and Spain; and part of the royal fifth was retained in America to pay for the colonies' own defence.[24] In an attempt to redress dropping trade flows and declining silver receipts, the Bourbon rulers of eighteenth-century Spain attempted to make the colonies profitable to the metropolis through the strict enforcement of mercantilist policies. The ultimate, though indirect, effect of their actions, as in the North American colonies, was to lose control of the region entirely.[25]

The internal markets of the Latin American colonies were not isolated from the rest of the world; rather, they were elements of a global system of exchange that evolved over the sixteenth to the eighteenth centuries. It is to a consideration of this system that we now turn.

The global system of exchange, 1500–1850

The global system of exchange that evolved between the fifteenth and the eighteenth centuries was structured by two broad tiers of trade. The first tier, intercontinental/oceanic trade was carried in European shipping and controlled, for the most part, by Europeans. The second tier, local trade, was carried out within and between independent producing regions and was controlled by locals and, where transported over water, was carried in local shipping.

The idea of 'locals' requires some explanation. In the Middle and Far East, this refers to non-Europeans. In the Latin American region, however, it refers to indigenous traders, Europeans born in America (that is, Creoles), and immigrant Spaniards who married locally and had strong reasons to invest in American trade. Indeed, the spectacularly rich 'great families' of Mexico, who based their fortunes on the vertical integration of production, manufacture and exchange in local, regional and inter-regional as well as international trade, were primarily Creoles and their Spanish nephews.[26] A rapidly increasing group of people of mixed descent also took part in local commerce as producers, manufacturers and, most especially, as traders. Members of this group moved to the fringes and frontiers of the empire on the Caribbean and Atlantic coasts in order to escape the difficulties associated with their ambiguous ethnic classification – and thus became some of the first to benefit when the Bourbons opened up American ports to freer trading at the end of the eighteenth century.[27]

The development of interoceanic trade dominated by the great companies was ultimately critical for global development and is by far the better-known aspect of global trade. However, Philip Curtin argues that the systems of local trade in the Far East and the Atlantic carried by far the greater volume of goods up to the middle of the nineteenth century.[28] And Patrick O'Brien, arguing against the claim by dependency theorists that European development was dependent on the accumulation of wealth through commerce with the periphery of a presumed global division of labour, writes that Europe actually gained very little in the way of profit before this date.[29]

If O'Brien is correct, then it is hard to see the rest of the world in terms of a 'periphery' in a global division of labour, with Europe at the centre. The notion that Europe benefited inordinately from intercontinental trade prior to 1850 was, O'Brien believes, a factor of documentation which has distracted attention from the activities of the greater part of the population:

> [O]ceanic trade has left an abundance of records which have seduced generations of historians eager to reconstruct the fascinating story of exploration, conquest, and rivalry among European states for the spoils of discovery. All that movement of men, ships, and exotic commodities, which attracted the attention of princes and became inseparable from the deployment of military power for political and economic ends, makes for readable history. Yet it is important to place the glamour of long-distance trade against the landscape of economic development. Braudel, who is the founding father of global history does just that in his classic study of the Mediterranean in the sixteenth century ... In his conclusion, Braudel ... emphasized how 'the Mediterranean in the sixteenth century was overwhelmingly a world of peasants and of the tenant farmers and landowners'; crops and harvests were vital matters of this world and anything else was superstructure, the result of accumulation and unnatural diversion towards the towns. Peasants and crops, in other words food supplies, and the size of the population silently determined the destiny of the age.[30]

In the case of Latin American historiography, attention was distracted from the realities of the colonial economies by the story of imperialism. When economic historians of the colonial era began to focus their attention on the political economies of the colonies as systems worthy of analysis in their own right – and not simply as economic dependencies of Europe – they uncovered evidence showing that the colonies produced more for the American markets than the European, and that, up to the last decades of the eighteenth century, these economies were oriented more towards the Pacific coast than the Atlantic. Indeed, Americans not only built the ships that carried their products up and down the Pacific coast of the Americas, they also controlled the (largely illegal) trade with Manila in the Philippines, and they built over 26 per cent of the Spanish merchant fleet carrying to the Americas between 1601 and 1650, and about 22 per cent between 1650–1700.[31] Ruggiero Romano demonstrates that, during the seventeenth century and up to 1740, far from being tied to European economic cycles, all sectors of the economies of Latin America and Europe were reversed: that is, when Europe was in crisis, Latin America was growing.[32]

Up to the end of the eighteenth century, the Latin American economies had an American rather than a European logic. This is not to say that the Latin American economies were miracles of development. By the end of the colonial era, they lagged far behind Europe in agricultural and transportation technologies. Comparative underdevelopment does not mean economic dependency, however, *unless there is a division of labour integrating the regions*. The Spanish Crown attempted to introduce just such a division of labour on its colonies through the imposition of monopolies, and failed miserably. Trade simply moved though contraband channels, and smugglers were aided and abetted by practically everyone, from the highest political officials down to the smallest petty traders.[33]

There is no doubt that, as the international systems of exchange grew, relations between states became increasingly shaped by considerations of international trade. But it is hard to sustain the argument that exchange at either the international or the local level was structured by a global division of labour: locals controlled local, regional and inter-regional trade; the 'centre' did not benefit to the extent predicted by the dependency model; and not only was the 'periphery' growing while the 'centre' was in crisis, but it enjoyed a degree of relative autonomy from the 'centre' that is incompatible with the dependency model. Discrete systems did exist within the international exchange system where regions were integrated by a division of labour: the triangular trade linking Europe, Africa and the plantations of the Americas (specifically, Brazil and the Caribbean and, later, the American South) is a well-known example.[34] But I would argue that the fact that we perceive these systems as discrete entities argues against a global division of labour: rather, they indicate what we *should* find everywhere, and do not.

We must ask, then, not only *when* and *why* Latin America became a peripheral resource region in a global division of labour, but when the global division of labour itself appeared.

Global transformations and local change

Up to the end of the eighteenth century, it can be argued that the evolving complexity of the international order was a function of the increasing amount of trade. But then something occurred in one specific region of this global system that transformed the international order: the replacement of animate with inanimate energy, creating the theoretical possibility of infinite increase in the production of manufactured goods. The development of the steam engine acted as a trigger to changes that cascaded through the international system over the next two centuries, transforming it into a system of production as well as exchange.[35] The process of industrialization itself was dependent to a great degree, though certainly not exclusively, on the prior presence of an international system. O'Brien argues that Europe contained sufficient resources to begin the processes of industrialization,[36] but without an international market and the resources of the colonial system that permitted access to raw materials, markets, and the possibility of capital formation far beyond what was possible within Europe, Britain would not have moved beyond a very local application of steam, as neither Britain, nor Europe for that matter, had the resources to sustain industrialization for very long. With the resources gathered by the international trading system, however, the process rapidly gained momentum; the slight edge of being the first to apply steam to production was translated into the flood of technological advances and social and political changes which carried Britain to world hegemony by the end of the nineteenth century.[37] In other words, industrialization developed within the context of an international system of exchange, and all the changes associated with industrialization, such as the appearance of capitalist relations of production and the formation of nation-states, fanned out through the system, resulting in the transformation of a system of exchange between producing regions into a system of distribution that sustained production in one part of the exchange network: that is, the global division of labour.

Over approximately the same period, the former colonies of Spain were first wrenched out of their Pacific solitude into the Atlantic economy and then integrated into the global division of labour as peripheral resource regions. The process began with the reform of the colonial system carried out by the Bourbon government of Spain over the last half of the eighteenth century. Through these reforms, the Bourbons attempted to make the colonies profitable for the metropolis in proper mercantilist fashion. They replaced Creoles with peninsular Spaniards in the government and the colonial Church; they increased the efficiency of tribute collection; and they attempted to increase the efficiency of the collection of tariffs. While these reforms clearly created a climate conducive to the independence movements, however, it was by opening up American ports to freer trading that the Bourbons succeeded in drawing the American economies into the European sphere. In Lockhart and Schwart's analysis, the proclamation of so-called 'free trade' in 1778 led to a reorientation of the American economies from the old colonial centres and the Pacific coast to the Atlantic coastal regions.[38] The development of vast expanses of land for livestock production, for wool, meat and hides, and the exploitation of the enormous natural reserves, which reached its culmination with the construction of railroads that shot straight to the heart of the continent, had begun.

Liberal ideology, classical economic theory and European products (and, later, finance), to say nothing of an increasing flow of military technology, cascaded out through the global exchange net that had evolved over the preceding 300 years. These novelties were embraced by local élites all over the world as providing them with the means to enhance their position both within local polities and within international trade, not always with happy results. The American colonies gained independence from their European metropoles and were transformed into nation-states, complete (in most cases) with liberal constitutions predicated on the notion of progress through production for export. They were also, in most cases, weighed down by debts inherited from the colonial era and exacerbated by the wars of independence from Spain. These debts were serviced by production for export and by recourse to the extraction of natural resources. In an attempt to increase the productivity of their lands, to progress through the application of new agricultural methods, Latin Americans extended commercial agriculture. They made use of a major tenet of liberal ideology – the notion that commons were antithetical to the protection of individual rights – to draw peasant village lands into commercial agriculture by denying the rights to commons that the Indians had enjoyed under the Spanish Crown. They also began to expand into the vast grasslands and to draw on their natural resources to an extent unknown in the colonial era.

As a result, Latin American environments were increasingly subject to homogenization and monocultural agricultural practices, with consequent loss of breeds and varieties, and were heavily exploited for raw materials, with consequent loss of biodiversity. At the same time, human populations increased exponentially, putting added stress on local environments. The increase in human population may simply reflect the rebound of the indigenous populations and the final playing out of the sixteenth-century process of biological integration; but the over-exploitation, and the associated destruction of natural resources, reflects the conjunction of local and global processes in the formation of a new global order.

The results for local environments – of vastly increased extraction of natural resources and the development of monoculture for the production of raw materials for export – are well known, if not particularly well studied. But one aspect of the process by which the Latin American states became structural elements in the rapidly evolving global system of production and exchange that has not been sufficiently taken into consideration is the ability of the new nation-states born of civil wars to pay their debts. We have been concentrating on the consequences of economic change for the environment in Latin American studies. Perhaps it is now time to look at the consequences of environmental change for economic development in the nineteenth century.

Notes

1. Robert B. Marks, 'Commercialization without Capitalism: Processes of Environmental Change in South China, 1550–1850', *Environmental History*, 1/1 (1996), pp. 56, 77.
2. Marks, 'Commercialization without Capitalism', p. 77.
3. Immanuel Wallerstein, *The Modern World-System: Capitalist Agriculture and the Origins of the European World-Economy in the Sixteenth Century* (New York: 1974).
4. For studies of the conquest era that did not take a self-conscious environmental history approach, but which have had significant impact on the development of a general perception of nature in the conquest era as victim, see Sherburne F. Cook and Woodrow Borah, *The Indian Population of Central Mexico, 1531–1610* (Berkeley: 1960); Lesley Byrd Simpson, *Exploitation of the Land in Sixteenth Century Mexico* (Berkeley: 1952). It should be noted, however, that these writers did not themselves treat the indigenous peoples as part of nature. For explicitly environmental history approaches to the conquest, see Alfred Crosby, *Ecological Imperialism: The Biological Expansion of Europe, 900–1900* (Cambridge: 1986); Elinor G. K. Melville, *A Plague of Sheep: Environmental Consequences of the Conquest of Mexico* (Cambridge: 1994). For studies of the colonial era that use European models, see Robert

II. Claxton, 'Weather-Based Hazards in Colonial Guatemala', *Studies in the Social Sciences*, 25 (1986), pp. 139–63; Murdo J. MacCleod, 'The Three Horsemen: Drought, Disease, Population and the Difficulties of 1726–27 in the Guadalajara Region', *Annals of the South-Eastern Council on Latin American Studies*, 14 (1983), pp. 33–47; Enrique Florescano, *Precios del maíz y crisis agrícolas en México (1708–1810)* (Mexico City: 1969).

5. Examples of studies which examine environmental reproduction include Kendall J. Brown, 'Mercury Mining at Huancavelica, Peru, and the Colonial Latin American Ecology', unpublished MS; Juan Carlos Garavaglia, 'Atlixco: L'Eau, les hommes et la terre dans une vallée mexicaine (15ème–17ème siècles)', *Annales*, 6 (1995), pp. 1309–49; Sonya Lipsett-Rivera, 'Puebla's Eighteenth Century Agrarian Decline: A New Perspective', *Hispanic American Historical Review*, 70 (1990), pp. 463–81; Ari Ouweneel, 'Schedules in Hacienda Agriculture', *Boletín de Estudios Latinoamericanos y del Caribe*, 40 (1986), pp. 63–97; Cynthia Radding, 'Ecology and Colonial Frontiers in Latin America', unpublished MS.

6. Carolyn Merchant, 'The Theoretical Structure of Ecological Revolutions', *Environmental Review*, 11/4 (1987), pp. 265–74; Fernand Braudel, 'Histoire de sciences sociales, la longue durée', *Annales: Economies, Sociétés, Civilisations* (1958), pp. 725–53.

7. See Warren Dean, *With Broadax and Firebrand: The Destruction of the Brazilian Atlantic Forest* (Berkeley: 1995).

8. Crosby, *Ecological Imperialism*; Dean, *With Broadax*.

9. Cook and Borah, *The Indian Population*; see also the *Latin American Population History Bulletin* for an ongoing debate over the size of pre-contact populations.

10. James Lockhart and Stuart B. Schwartz, *Early Latin America: A History of Colonial Spanish America and Brazil* (Cambridge: 1983).

11. For a comparison of the Inca and Aztec Empires, see George A. Collier, Renato I. Rosaldo and John D. Wirth, *The Inca and Aztec States, 1400–1800: Anthropology and History* (New York: 1982).

12. Charles Gibson, *The Aztecs under Spanish Rule: A History of the Indians of the Valley of Mexico, 1519–1810* (Stanford: 1967).

13. The *conquistador-encomenderos* received the right to labour and tribute, but not to land.

14. Gibson, *Aztecs*, p. 322.

15. Arnold J. Bauer, 'La Cultura mediterranea en las condiciones del Nuevo Mundo: Elementos en la transferencia del trigo a las Indias', *Historia*, 21 (1986), pp. 31–53.

16. For a discussion of the introduction of specific cattle breeds into Mexico and Texas, see William Doolittle, 'Las Marismas to Pánuco to Texas: The Transfer of Open Range Cattle Ranching from Ibéria through North-Eastern Mexico', *Yearbook: Conference of Latin Americanist Geographers*, 13 (1987), pp. 3–11.

17. Gibson, *Aztecs*, pp. 322–3.

18. Steve J. Stern, *Peru's Indian Peoples and the Challenge of the Spanish Conquest* (Madison: 1993), p. 37.

19. Bernadino de Sahagún, *Florentine Codex: General History of the Things of New Spain*, trans. Arthur J. O. Anderson and Charles E. Dibble (Salt Lake City: 1950–82).

20. Stern, *Peru's Indian Peoples*, p. 38.

21. Although he is writing of the mid- to late colonial era, Robert Patch describes this system very clearly: 'Imperial Politics and Local Economy in Colonial Central America, 1670–1770', *Past and Present*, 143 (1996) pp. 77–107.

22. Patch, 'Imperial Politics', p. 105.

23. Lockhart and Schwartz, *Early Latin America*, p. 153.

24. Peter J. Bakewell, *Silver Mining and Society in Colonial Mexico: Zacatecas 1546–1700* (Cambridge: 1971), p. 235.

25. Ruggiero Romano writes that, '[d]uring the eighteenth century the recovery of colonial affairs (for America, the Spanish reforms meant just that) put an end to a relative autonomy. However, the tensions which arose were so strong they resulted in the final independence of the colonies': *La Crise du XVIIe siècle* (Geneva: 1992), p. 232.

26. Lockhart and Schwartz, *Early Latin America*, pp. 153–4, discusses the development of American ties among the great merchants; see also John E. Kicza, *Colonial Entrepreneurs, Families and Business in Bourbon Mexicao City* (Albuquerque: 1983).

27. Lockhart and Schwartz, *Early Latin America*.

28. Philip Curtin, *Cross-Cultural Trade in World History* (Cambridge: 1990), p. 156.

29. Patrick O'Brien, 'European Economic Development: The Contribution of the Periphery', *The Economic History Review*, 2nd Series, 35/1 (1982), pp. 16–17.

30. O'Brien, 'European Economic Development', p. 18.

31. Richard J. Salvucci (ed.), *Latin America and the World Economy: Dependency and Beyond* (London: 1996). Patch, 'Imperial Politics', found that the 'production [of textiles by the indigenous Maya] made Central America into an export platform which, like India and China, sent its textiles not to the core of the European economy but rather to other regional economies of Latin America' (p. 106). Figures for American ship building cited in Romano, *La Crise*, p. 230.

32. Romano, *La Crise*, p. 231.

33. Smuggling is extremely difficult to study, but scholars agree that it was very extensive in the Caribbean and Pacific; see, for example, Stanley J. and Barbara H. Stein, *The Colonial Heritage of Latin America: Essays on Economic Dependence in Perspective* (New York: 1972), esp. pp. 50–3. On the Atlantic, see Philip D. Curtin, *The Rise and Fall of the Plantation Complex: Essays in Atlantic History* (Cambridge: 1990), p. 132. For difficulties in studying smuggling, see Carlos D. Malamud, 'El comercio directo de Europa con América en el siglo XVII: Algunas consideraciones', *Quinto Centenario* (1981), pp. 25–52.

34. The plantation complex of the South Atlantic would seem to provide a clear example of colonial production for the benefit of the metropolitan regions, but even here we find the rapid formation of an extensive smuggling network: see Curtin, *The Rise and Fall of the Plantation Complex*, p. 132.

35. See Paul Kennedy, *The Rise and Fall of the Great Powers: Economic Change and Military Conflict from 1500 to 2000* (London: 1989), esp. pp. 189–203.

36. O'Brien, 'European Economic Development', p. 7.

37. Kennedy, *Rise and Fall of the Great Powers*, pp. 189–203.

38. Lockhart and Schwartz, *Early Latin America*, chapters 9–11.

CHAPTER 13

The Transvaal beef frontier: environment, markets and the ideology of development, 1902–1942

Shaun Milton

The development of settler cattle and beef production in South Africa from the end of the South African War (1899–1902) to the outbreak of World War II was shaped by the influences of war, nature, imperialism, undercapitalized settler colonialism and, by extension, the limited technical, material and political resources of the settler state. More directly, the expansion of settler cattle production was part of the institution and extension of colonial power into the rural areas of the Transvaal after the end of the South African War.[1]

At this time, the Transvaal's hinterland, particularly the western Bushveld to the north and west of the Witwatersrand, was a zone of potential colonization within a pre-existing set of recognized international boundaries, on what could be defined as a frontier of secondary settlement.[2] Unlike the history of rural transformation in the settler colonies of 'demographic takeover', such as Australia, the Transvaal had an indigenous population that far outnumbered the colonizers.[3] Thus the need to assert colonial authority on this frontier was more pressing than, say, in northern Queensland. The new colonial state attempted to assert its authority spatially by altering the rural demographics and by placing as many white agricultural settlers as possible on the land under the so-called 'European land settlement schemes'. The schemes were, in turn, supported by the extension of the ecological and administrative frontiers through the application of veterinary resources and the 'paraphernalia' of surveying, fencing, and beacons.[4]

As most of the land available to the settlement schemes in the Transvaal lay in the semi-arid areas of the western Bushveld, cattle production was deemed the most viable agricultural activity for 'opening up' these areas. Thus, a new 'beef frontier' was established which was intended to underpin this frontier economically by providing cheap beef to the local urban mine-dominated markets.

The settlement projects in the Transvaal were also part of a broader policy of 'encouraging' Africans to take up wage labour in the recovering settler economy. The colonial state went to great lengths to marginalize and pauperize African rural societies through land dispossession and restrictions on peasant production, including protectionist measures against a competing pre-colonial tradition of cattle husbandry and trade.[5] Settler cattlemen were more concerned

with appropriating rangeland and beef markets than labour, but they nevertheless played an important, if less direct, role in this process. In this context,
settler-owned cattle became the 'shock troops' of colonial consolidation.

The ideology of settlement and 'progressive' agriculture

After 1902, imperial and settler interests in the Transvaal gradually harmonized
around a desire to re-establish and maintain the viability of gold production
on the Rand. In turn, gold production (or at least its perceived potential)
underpinned the establishment and development of the settler colony in the
Transvaal and, later, in the Union after 1910.

Alfred Milner, Governor of the Transvaal, 1901–5, was the chief architect of
the Transvaal's postwar reconstruction, which included a rural development
programme. The justification for this project was articulated through imperial
and positivist ideas of 'progressive' agriculture and estate management, and racial
supremacy. The most important of these was a belief in the superiority of whites
over Africans. From this, imperial and later settler administrations deduced that
the political requisite for the future economic and social development of the
Transvaal – and, within this, the securing of long-term colonial interests – was
for the colony to become a 'white man's country'. The result of this dispensation
was, of course, the subordination of African interests to those of the white
settlers.[6] This was why, in 1902, the British did not entrust the local administration of the Transvaal hinterland to the African chieftaincies, or encourage
African peasants to produce crops and livestock for the local mine-dominated
market. Not only did such an option lie outside the bounds of what was practical
in settler politics, but an economically viable peasant sector might have disrupted
the flow of cheap African labour to the mining industry.

European land settlement schemes and settler-based agriculture naturally
favoured settler production over African production. Apart from racial supre-
macist assumptions, settler production was justified by a kind of 'agricultural
chauvinism': a belief in the primacy of western 'science-based' production
methods, private property, and agrarian romanticism. It was an outlook that
combined notions of 'nostalgic pastoralism' and 'agricultural improvement'
with commercialization, to form what Paul Rich has termed a kind of 'technological pastoralism'. In both Australia and South Africa, evangelists of this
ideology dreamt of a 'sturdy yeomanry' settled on deeded estates, practising
agriculture so sustainable that these pioneers would be able to 'recreate [green]
England in a [brown] land of exiles'.[7]

Initially, the British envisaged the installation of an Anglo 'yeoman' farmer
class to 'set a good example' to other farmers and to be an important 'link
between the Government and the [returning] Dutch rural population'.[8] In later
years, this stratum came to include farmers who simply considered themselves
'progressive', including many Afrikaners. A shared 'progressive' ethos provided the ideological basis for an abiding, if not always easy, alliance between

technocrats from the Department of Agriculture and a small but influential group of Afrikaner and English-speaking farmers and landowning plutocrats.[9] Many in the farming élite were cattle producers and it was cattlemen who came to dominate what became the bastion of 'progressive' farming in the Transvaal, the Transvaal Agricultural Union (TAU). During the period under review, the leadership of the TAU and, later, the South African Agricultural Union was dominated by mainly English-speaking, Transvaal cattle farmers. For most of the first four decades of this century, the Transvaalers were led by two Australian soldier-settlers: Major R. D. Doyle and Major E. W. Hunt.[10]

The influence of this farming élite on the formulation and implementation of agricultural policy cannot be overestimated. Even after the defeat of the South African party (the party of the 'progressives') in 1924, the long-established and close working relationship between prominent 'progressive' cattle farmers in the TAU and senior officials of the Department of Agriculture continued.[11] The power of the 'progressives' reached its zenith with the establishment of the Meat Board in 1934, which basically gave them the task of regulating their own industry through a system of quotas.[12]

The original political function of 'progressive' settlers on the land and of settler agriculture continued to be acknowledged by at least a part of the British imperial élite as a crucial ingredient in the maintenance of colonial power. In July 1925, at a dinner given at the Hyde Park Hotel in honour of a group of visiting Union farmers, the Colonial Secretary, Leo Amery, in proposing a toast to Milner's agricultural legacy, asserted that '[settler] Agriculture would help to keep South Africa white because the proportion of whites to blacks was far higher on the land than it was in the deep levels of the mines'.[13] It was a role that many 'progressives' saw themselves as playing on the Transvaal beef frontier. In a letter to the *Farmers' Weekly* in 1918, one prominent farmer went so far as to suggest that 'our future depends on the occupation of the land by the white race ... [through] the development of its natural pastoral wealth by stock-farming'.[14]

Between 1906 and 1910, half of all the settlement land disposed of annually in South Africa was located in the western Bushveld; this proportion rose to about 60 per cent after 1914, where it remained well into the 1930s. In some districts of the western Bushveld, herds increased by between 35 and 73 per cent.[15] The steady flow of 'progressive' settler cattle farmers to the western Bushveld during the first four decades of this century is a testament to the powerful attraction of this settler ideal.[16] Yet many of the rank-and-file unionized farmers in drought- or locust-stricken areas like the western Bushveld were undercapitalized, debt-ridden 'Land Department settlers'. They hardly fitted the 'progressive' paradigm. In these extreme circumstances, there was probably little to distinguish them from their more 'traditional' neighbours. As one farmer put it, it was much easier to preach 'progressive' farming than to practise it. Publicly expressed loyalty to these ideas, however, provided advocates with a social or political badge that distinguished them from those they considered to be their social inferiors.[17] White 'progressive' farmers were thus the local guardians and trustees of an expanding racialized moral economy, based on deeded property and profit, more vigorous and extensive than anything that had been possible under the old (pre-South African War) republican regime.[18]

Systemic contradictions

As early as 1906, the Transvaal's Secretary for Agriculture, F. B. Smith, rightly surmised that there was 'some deep seated cause' that was arresting the development of the Transvaal beef frontier. He was coming to realize that the root of the problem lay in a contradiction in state policy, in essence between the 'politics' of white settlement and its 'economics'.[19] This contradiction was intersected by three perennial and interrelated factors which explain why settler beef production in the western Bushveld remained a risky business over the coming decades.

Size and price of settler allotments in a semi-arid zone

The savanna environment of most of the Transvaal beef frontier was characterized by inadequate rainfall and a low carrying capacity. Early critics of Milner's settlement policy were quick to recognize that the size of the settlement farms was not economically viable.[20] They recommended that agricultural and pastoral production should be the preserve of the African population, who were already acquainted with the marginal qualities of the soil and the climate. To 'force' agricultural development, they warned: 'in order to satisfy a certain political expediency [contained within the settlement schemes] must result in the creation of a class of agricultural paupers which must ultimately prove a source of serious weakness to the Government of the future'.[21] In addition to the political dangers of such an outcome, the critics warned, there was also a risk of overproduction of beef. But the colonizing imperatives of land settlement were such that the administration pressed ahead regardless.[22] This contradiction continued to manifest itself in the differences between the aims of the Departments of Lands and Agriculture. Although certain 'progressive' settlers and agricultural officials were acknowledging by the 1920s that 18,000 morgen (1 morgen equals 2 acres or 0.8 hectare) was the minimum size for viable stock farming, most government allotments in the western Bushveld were still as little as 1,400 and 2,000 morgen with a significant number of only 480–700 morgen in extent.[23]

Another major ecological drawback of the beef frontier was that large parts of it were covered by 'sweet veld'. Although this is good winter pasture, it has serious nutritional deficiencies. With the decline, after 1902, of more informal systems of land tenure like free commonage, seasonal transhumance became more difficult and herds were increasingly confined to the same farm all year round. Apart from increasing the likelihood of overgrazing, the lack of access to summer pastures or supplementary feeding increased the time that it took to produce a steer for market. This added to the overall production costs.[24] Further natural hindrances included a high susceptibility to droughts and locusts. Between 1902 and 1942, the western Bushveld was hit by almost every major drought in South Africa, the worst occurring in the early 1930s, which in turn contributed to the accumulating legacy of socio-economic decline.[25]

The economics of deep-level, low-grade gold production on the Rand provided the other factor profoundly affecting the way in which settler beef production subsequently developed in the new Transvaal cattle-raising areas. Significantly, it encouraged the overvaluing of land. Many farms were bought and sold with at least part of the price based on their mineral prospects rather than on their agricultural potential alone. Even land released under government schemes was required to reflect prospecting prices. The farming of overpriced semi-arid estates that were too small to be economically viable resulted in low-grade expensively produced settler cattle.[26]

The gold mines and their domination of the local markets

It was originally intended that the new beef frontier would find a market in the anticipated expansion of gold production. But successive depressions and the erratic expansion of gold mining on the Rand tended to preclude long-term stability or to provide the sustained and increased demand for beef to match the increases in cattle production, at least until the mid-1930s. Moreover, when increasing numbers of Transvaal settler-produced cattle did start to come on to the Rand market after about 1906, cheaper African-raised stock began to compete successfully for new market opportunities.[27]

The extraction of deep-level, low-grade ore was expensive, requiring the mines to work on very narrow profit margins, so ration beef was sought at the cheapest prices with little regard for quality. Gold-mining houses were the largest buyers of beef in South Africa, the Johannesburg price for low-grade 'compound' cattle providing a benchmark that tended to dampen average wholesale prices throughout South Africa.[28] The influence of the Johannesburg market increased after the unification of South Africa in 1910, as the economies of the four former colonies became more integrated with the removal of inter-colonial customs and tariff barriers.

The mines came to rely on the cheaper supply of African-produced cattle from the African reserves of the eastern Zoutpansberg, Transkei, Swaziland, and the Bechuanaland Protectorate. African stock was able to compete with settler-produced cattle from the western Bushveld largely because of communal land practices. But the effects of colonial conquest and consolidation, including certain protectionist marketing and veterinary arrangements, also forced desperate African producers to accept low wholesale prices. After 1911, African-produced cattle were invariably sent to the low-priced 'compound' market in Johannesburg, often via something called the 'quarantine' auction floor. Here they were joined by settler- and ranching company-produced cattle from Southern Rhodesia and – after World War I – South-West Africa, where land prices were lower and settler holdings were more extensive than in most parts of South Africa.[29] Stock producers outside the South African settler sector were better placed to absorb the lower prices on what rapidly became, and largely remained, an oversupplied market on the Rand.

Thus, chronically low wholesale prices discouraged settler investment in breeding stock, fencing, wells, and pasture improvement, particularly in the semi-arid western Bushveld. In addition, foreign investment in South Africa during this period tended to be placed in mining or mineral land speculation, rather than in agricultural or livestock production. Certainly, for the large metropolitan meat companies, ranching or meat processing in the wilds of the north and west Transvaal could not match the ecological and geographical attractions of beef production on the fertile soils of the valley of the Plate River.[30] In addition, low taxes on mining revenues limited the resources available for investment in the beef industry.[31]

The problems of science and science-based resources for settlers

The expansion of the fledgling industry and white settlement on the beef frontier was greatly facilitated by the successful deployment of new veterinary resources and regulation, particularly in arresting the spread of East Coast fever. This partial 'taming' of the environment generated a further wave of settler migration to the north and, with it, an unprecedented and sustained expansion of the Transvaal settler-owned cattle population. However, this expansion was not attended by a comparable improvement in the quality of beef and the cost of production or transport and processing infrastructure. While the successful control of bovine disease assisted in the process of rural colonial consolidation, the unintended biological consequence of this narrow application of state resources was to contribute to the generation of a regional surplus in low-grade cattle.

Settler beef producers in the Union were subjected to the bureaucracy of science through veterinary regulation. For potential 'progressives', this experience, along with forms of written propaganda, contributed to their knowledge of science and its attendant technology, all of which informed their overall agricultural ideology and developmental intent. However, inadequacy or absence of other state extension services and the lack of any local commercial incentive or opportunity for settlers to provide such input themselves meant that white farmers were often bereft of the material resources to apply this scientific knowledge to the production process effectively and comprehensively, in a way that might have given them a competitive advantage.[32]

The structural factors associated with settler production on the Transvaal beef frontier encouraged the evolution of an export-led development strategy for the cattle industry. Although it was ostensibly driven by the need to find solutions to overproduction and outside competition, at its root was the necessity to persuade whites – and 'progressive' settler farmers in particular – to remain in the remote parts of the beef frontier.[33]

In what would become almost an obsession, the Department of Agriculture's search for an export market can be ascribed to a settler culture or 'mentality'. Exports appealed to a positivist view of national development and were considered an important expression of national self-assertion. It was a view that was shared with white settlers around the world. Beef exports would bolster the Union's political image as a 'white man's country', assertively fulfilling its 'manifest destiny' not only in its own hinterland, but as a member of the worldwide community of 'white' dominions with a seat at the imperial high table.[34] Additional encouragement to develop a beef export industry came from an imperial government largely motivated by periodic concerns over food security and the home market's over-reliance on Argentinian beef.[35]

Early efforts to export beef failed. When measured against the standards required on the international market, South African beef (especially from the western Bushveld) was considered too poor in quality and too expensive.[36] It was only with the outbreak of World War I, and the higher demand and prices for beef that this generated, that there was any sense that the earlier promise of prosperity might be fulfilled and that 'progressive' settler farmers would be emancipated from overproduction, African competition, and the stifling domination of the mine-compound trade.[37] The war generated an unprecedented demand for South African beef from the imperial army. After some delay due to a prolonged drought, the Union was exporting army beef in significant quantities by 1916, mainly to the Middle East. Prices were pushed high enough to offset temporarily the structural impediments associated with the cattle industry. The high prices, in turn, encouraged a further expansion in white settlement and settler beef production. There was a concomitant rush of speculative investment in land and plant.[38]

With the predictable postwar slump to prewar demand levels, when British demand switched back to chilled rather than frozen beef, the shallow foundations on which the expansion of this industry had been based were revealed. If anything, the pre-existing contradictions and shortcomings of the industry were magnified. The settler industry was now structurally dependent on an export market that no longer existed. The only substantive change to be brought about during the war years was an increase in the capacity of the region to produce a commodity of which it already had an abundance – low-grade slaughter stock.[39] Settler-produced beef originally intended for export was now diverted to the Rand, where it had to compete again with African and ranching stock from neighbouring territories.[40]

A new export market for frozen beef did emerge in the 1920s: the Italian army in Libya and Eritrea. However, Imperial Cold Storage, the South African meat company, won this contract by basing its bid on its more competitive cattle and land assets outside the Union. While capital investment and beef exports in the Union remained stagnant, Imperial Cold Storage went on to develop the Italian trade by erecting new meat plants and acquiring additional ranches in Rhodesia, Bechuanaland and South-West Africa.[41] It was only in

1939 that South Africa was able to establish a modest, but steady, trade in chilled beef with Britain. With the outbreak of World War II, however, international demand again switched back to frozen beef, and domestic demand pushed wholesale prices up still further.[42]

The racial and class dimensions of protectionism and 'quality control'

For most of the 1920s and 1930s, the Union's efforts to protect and promote white cattle producers were largely shaped by the consequences of the industry's wartime expansion. 'Progressive' farmers, particularly in the TAU, led a loose alliance of Union settler producers and Afrikaner nationalists to demand that the government restore domestic wholesale prices by imposing a total embargo on cattle imports. A total embargo was never really an option for the settler state, because of the reliance of the mines on cheap imported cattle. However, by way of compromise, a string of market and weight restrictions were imposed between 1923 and 1934.[43] Every new measure that was added to this partial embargo had an opposite and incremental effect to its intended purpose: the wholesale prices of imported cattle were driven down – thereby extending the domination of non-Union cattle in the mine-compound trade – while the stifling of legitimate trade encouraged increasing levels of cross-border cattle smuggling from the Bechuanaland Protectorate and Southern Rhodesia.[44]

Economic protection became tangled up with issues of cross-border veterinary protection. The expansion of mining and industrial production after 1934 pushed the demand for beef on the Rand to even higher levels, much of which was met through the underground smuggling network. In response to the demands of Bushveld 'progressive' farmers in the TAU, the government passed the anti-cattle-smuggling amendment to the Livestock Diseases Act in 1937, which was a trade protectionist measure passed off as a veterinary regulation.[45] Attempts to limit smuggling were never sustainable, however, invariably hampered by limited material and personnel resources and the remoteness of the beef frontier.

At the same time, the settler state pursued a policy that sought to 'Europeanize' cattle production under the guise of 'upgrading'. Manifestations of this racial intent are apparent in the differing respective policies of the Departments of Agriculture and Native Affairs. In the case of settler-owned cattle, 'improvement' was envisaged as persuading settlers to 'upgrade' their productive capacities through cross-breeding and 'progressive' production methods, thereby improving the quality and competitiveness of their cattle in order to meet the standards required for export.[46] By contrast, 'upgrading' for African-owned cattle was largely intended to be through stock reduction and castration.[47] The vested interests and rivalry of the established and powerful breeding societies, coupled with a kind of bovine racism, undermined the inquiry into the use of Afrikaner and indigenous (rather than exotic) breeds in semi-arid settler areas.[48] Efforts to reduce African-owned herds floundered, due to the reluctance of local officials and chiefs to enforce what they saw as unenforceable de-stocking proclamations.[49] For most of the 1920s, tax-raising legislation to support 'upgrading' was blocked by anti-farming nationalists inside the Union Parliament. But the passing of the Livestock and Meat Industry Act in 1934 brought some semblance of regulation and organization to the settler cattle industry. Contained within the law was provision for a Livestock and Meat Board, empowered to raise funds through a slaughter levy in order to subsidize exports. Provision was also made in the Act to encourage the improvement of quality through measures such as the establishment of settler 'cattle improvement areas'.[50]

For settler 'progressives' in the borderlands of the western Bushveld, breeding stock, fencing, boreholes, dams and watered pastures did not come cheap in the face of drought, high-interest loans, and low market prices. Beyond the 'dreaded border' with the Bechuanaland Protectorate lay what they perceived as the ultimate threat: the unregulated, disease-ridden, backward, but above all cheap and, therefore, potentially competitive, Bechuana African cattle economy. It seemed 'unfair' to these settlers that progress and improvement should be thwarted by the illegal introduction of cattle from these areas by people, both black and white, whom they considered, to be locked into the old ways.[51]

The cardinal transgression of white cattle smugglers, many of them *arme grens mense* (poor border people), was their willingness to cross the racially defined boundaries of production that had been imposed after 1902 to trade directly with African producers in the Protectorate. These 'freebooters' were regarded by their 'progressive' neighbours as treacherous and degenerate, as the very harbingers of a kind of bovine *swart gevaar* (black threat), which, in this context, was a mix of fears associated with bovine infection and African competition – for markets, pasture and water resources. In short, the activities of smugglers subverted the modernist ideals of 'progressive farming' and undermined the racially ordered colonial society.[52]

Ironically, in the long term, it was not exports but the massive industrialization and urbanization of the Witwatersrand during World War II and then the extra revenues and markets provided by new gold fields in the early 1950s that finally delivered some measure of sustainability to settler producers as South Africa entered the apartheid era. Thus the vision of the yeoman farmer was reaffirmed. Exports were largely abandoned, but most of the other elements of settler beef production policy remained and served to support the political aims of apartheid. African producers continued to be marginalized as a more systematic form of social and economic segregation was attempted.

Policy reassessed

In spite of the wartime boom conditions prevailing by the early 1940s, the experience of the 1920s had taught the agricultural planners that questions of agricultural and land sustainability needed to be addressed before peace was restored. On the subject of white settlement, opinion was divided. The settler state could rightly claim in 1939 that, in spite of the difficulties in encouraging production, the minimum rural (political) objective had been largely fulfilled during the intervening years: its policies had kept 'the [white] farming population on their farms', thereby preserving a white presence in the countryside, particularly in the more remote areas like the western Bushveld.[53] However, the viability or necessity of this approach and its embodiment in the European land settlement schemes were now being called into question.[54] In fact, strong doubts were being expressed in some quarters about the 'practicability of [agricultural] economic segregation'.[55] But the continued existence of African-owned cattle was a reminder of the settlers' own historically precarious position.[56] The following extract from a letter to the *Farmers' Weekly* in 1939 not only reflects the failure of the settler state to complete its project of promoting white interests over those of Africans in the cattle trade, but also the possible implications of that failure:

> Natives own more cattle in the Union than Whites ... [A]ided by the adjoining Native territories, Natives can supply the beef requirements of the Union ... which recalls the Chamber of Mines suggestion, that in order to obtain cheap agricultural products all agriculture should be left to the Natives. If this was to eventuate in this primarily stock raising country, then the majority of White farmers would be driven off the land ... [The] Union must decide if it is essential to the preservation of the White race to have a White rural population living under civilized conditions.[57]

By the early 1940s, the Departments of Agriculture and Lands were becoming more alive to the contradictions inherent in the, by now, forty-year-old project of combining settler agricultural production with white settlement as a way of colonizing remote marginal areas like the western Bushveld. The 1943 *Report on the Reconstruction of Agriculture* appeared to be testing the water for at least a partial discontinuance of the project. The report's recommendations included the abandonment of the export cattle programme, with the recognition that the 'natural limitations' of the Union made such a course environmentally unsustainable for the foreseeable future. In semi-arid areas like the western Bushveld, where farming was 'hazardous' as well as 'socially undesirable', a 'partial' re-nationalization of land was envisaged in order to limit speculative price rises and to encourage more efficient and extensive livestock production. Special 'reserve grazing areas' were proposed for settler production, where land tenure would be restricted to larger, more economically viable ranches on long leases (after the Australian model), and only on condition that, as the report put it, strict 'progressive' guidelines on production and land usage were followed. In addition, the report recommended the nationalization of all meat processing in the Union.[58] The report also gave belated recognition to the importance of Union and non-Union African-produced cattle to the dominant sectors of the settler economy, and the necessity for preserving and developing these resources. All these recommendations signalled at least the beginnings of a re-evaluation of the settler yeoman ideal, and can be seen as part of a brief liberal moment in South Africa.[59]

Conclusion

During the period 1902–42, neither the South African state nor its antecedent in the Transvaal possessed the political or material resources to reconstitute, in its own ideological image, the social, economic and environmental conditions pertaining to the Transvaal beef frontier or the beef cattle industry generally. The impact of state intervention was, however, sufficiently disruptive to generate a set of unintended consequences which became characteristics of the cattle industry. The fundamental flaw in state policy was that the political imperative of asserting colonial authority through a racial demarcation of the landscape took precedence over the development of a commercially viable cattle industry at ease with its human and physical environment. The deficit between ideological intent and actual achievement through rural development is illustrated by Milner's visit to the Transvaal western Bushveld in December 1924 to view his legacy, nearly twenty years after he had left. Far from seeing the wilds tamed, 'the father of the settlers' caught sleeping sickness there, took ill on his way back to Britain, and died at his home in May 1925.[60]

The enlightened policies instituted towards African cattle production during the last years of the second Smuts government were reversed or significantly modified after the National party victory in 1948. The change in government provided the necessary political will to effect a restoration and expansion of central aspects of earlier policies to promote settler cattle production.[61] Moreover, with the opening of the new Orange Free State gold fields in the early 1950s came a significant increase in state revenues. The Department of Agriculture was now able to deploy far greater resources, to a wider set of recipients, in support of these and other agricultural policies than had ever been possible in the preceding decades.

As late as the 1980s, however, white cattle farmers in the western Bushveld continued to face environmental, infrastructural, and marketing problems similar to those that had met the first wave of settlers after the South African War. There was some consolidation of land ownership, but most settlers still attempted production on farms that were too small to be economically viable. In addition, security and state subsidy and support were draining away as the economic and political power of settler South Africa declined. Many of the old farms in the western Bushveld were abandoned by whites, thus removing the key motive for introducing commercial cattle production to the area in the first place. Other ranches gave up cattle production and became game farms.[62]

South Africa now needs to provide sufficient food to feed its rapidly expanding population. In the debate about sustainable land use in South Africa it will be important for the government to resist any temptation to allow short-term political or economic considerations to drive rural development programmes, to the detriment of longer-term considerations of productive and environmental sustainability, or for policy to be influenced by the discredited agricultural and scientific ideologies of old.[63] If the mistakes of the past are to be avoided, then hard questions must be asked about how livestock production is to be organized in the future and under what form or forms of land tenure it should be conducted. In this respect, the history of beef production in the western Bushveld can provide crucial lessons for the future.

Notes

1. Milner (Governor), Johannesburg, to Chamberlain (Secretary of State), London, 'Relief and Resettlement', 25 April 1902 (Public Records Office [PRO]: Colonial Office [CO] 417/351 19491).

2. J. R. V. Prescott, *Political Frontiers and Boundaries* (Sydney: 1987), pp. 36–7, 41.

3. Melbourne (Governor), Johannesburg, to Secretary of State, London, Confidential, 'Land Settlement', 31 July 1905 (PRO CO291/83 29962).

4. I. Hofmeyr, *'We Spend our Years as a Tale Told': Oral Historical Narrative in a South African Chiefdom* (Johannesburg: 1994), pp. 59–77.

5. *Annual Report of the Native Affairs Department to March 1903*, pp. A10, A17, A23 (Transvaal Government [TG]).

6. Selborne, Johannesburg, to Elgin, London, 'Native Policy', 13 January 1908 (PRO CO291/125 3907).

7. Paul Rich, 'Milnerism and a Ripping Yarn: Transvaal Land Settlement and John Buchan's Novel *Prestor John*, 1901–10', in B. Bozzoli (ed.), *Town and Countryside in the Transvaal* (Johannesburg: 1983), pp. 414–18; M. Adas, *Machines as the Measure of Man: Science, Technology, and Ideologies of Western Dominance* (Ithaca: 1992), p. 404; Neil Barr and John Cary, *Greening a Brown Land* (Melbourne: 1992), p. 3.

8. Selborne, Johannesburg, to Lyttelton (Secretary of State), London, Confidential, 'Land Settlement', 31 July 1905 (PRO CO291/83 29962); I. Hofmeyr, 'Turning Region into Narrative: English Storytelling in the Waterberg', in P. Bonner (ed.), *Holding their Ground: Class, Locality and Culture in 19th and 20th Century South Africa* (Johannesburg: 1989), pp. 268–9.

9. A. J. Christopher, *Crown Lands of South Africa* (Kingston: 1984), pp. 156–9, 170–8.

10. 'Doyle, Mjr. Richard Dines', *South African Who's Who: 1935* (Johannesburg: 1935), p. 56; *Farmers' Weekly*, 15 November 1933, p. 534.

11. *Farmers' Weekly*, 24 September 1924, p. 151.

12. Gov. Sec to A. Miller Mbabane, 29 December 1936 (MS 719, Alister Miller Papers, Killie Campbell Library, University of Natal [KCL]); *Farmers' Weekly*, 10 April 1935, p. 303.

13. *Star*, 4 July, 31 July 1925.

14. *Farmers' Weekly*, 17 July 1918, p. 2183.

15. 'Comparative Agricultural Statistics for the Transvaal 1908-10', *Agricultural Journal of the Union of South Africa*, 1/2 (1911), p. 281; *Annual Report of the Department of Lands, for the Year Ended 31 March 1922*, pp. 379–83 (Union Government [UG] 14-23); *Annual Report of the Department of Lands, for the Year Ended March 1927*, pp. 25–8 (UG17-28).

16. *Report of the Department of Lands, 1922*, pp. 379–83; *Report of the Department of Lands, 1927*, pp. 25–8; *Farmers' Weekly*, 10 September 1924, p. 2655.

17. *Farmers' Weekly*, 2 June 1926, p. 1176; 5 December 1928, p. 1269.

18. Rich, 'Milnerism and a Ripping Yarn', p. 421; M. Cowen and R. Shenton, *Doctrines of Development* (London: 1996), p. 363.

19. *Annual Report of the Department of Agriculture, 1905–6*, pp. 58–61 (TG).

20. Rainfall is unreliable and occurs in summer in the form of thunderstorms; in a good year, 35 to 65 cms can be recorded: J. S. P. Naude, 'Live Stock Farming in the Northern Transvaal,

1927–30, Part 2', *Department of Agriculture: Economic Series Bulletin*, 129 (1934), p. 7; A. J. Christopher, *The World's Landscapes: South Africa* (Harlow: 1982), p. 90.

21. Milner, Johannesburg, to Chamberlain (Secretary of State), London, Confidential, 'Agricultural Prospects – re. Report by Lt. Col. Owen Thomas to the TCLE Co', 16 March 1903 (PRO CO291/55 12658).

22. Naude, 'Live Stock Farming', pp. 39–41; *Report of the Land Settlement Commission, South Africa*, BPP Cd 626 (London: 1901).

23. *Farmers' Weekly*, 27 October 1920; *Report of the Department of Lands, 1927*, pp. 23–41; *Annual Report of Department of Lands, for the Year Ended March 1939*, p. 16 (UG23-40); A. M. Bosman, 'The Beef Industry in South Africa', *Farming in South Africa*, February 1924, p. 212; P. J. v.d. Schreuder, 'Breeding for Beef', *Journal of the Department of Agriculture*, January 1923, p. 29; Naude, 'Live Stock Farming', pp. 39–41.

24. Lack of phosphorous was significant: Christopher, *The World's Landscapes*, pp. 12–15; A. Smith, *Pastoralism in Africa: Origins and Development Ecology* (Johannesburg: 1992), p. 135; Bosman, 'The Beef Industry', p. 202. For veterinary considerations of kraaling and transhumance, see William Beinart, Chapter 6, above.

25. Droughts occurred in 1908, 1912–15, 1916, 1919, 1926–7, 1932, 1938: L. Levenkind, 'Droughts in South Africa', *Farming in South Africa*, March 1941, p. 87; *Farmers' Weekly*, 11 July 1934, p. 1113; *Annual Report of the Department of Agriculture, 1933*, reprinted in *Farming in South Africa*, December 1933, pp. 448–50.

26. R. Morrell, 'Farmers, Randlords and the South African State: Confrontation in the Witwatersrand Beef Markets, c.1920–23', *Journal of African History*, 27 (1986), p. 516; Magistrate Vryburg to Sec. for PM, Pretoria, 23 October 1922 (South African State Archives [SASA]: Department of Agriculture [LDB] 1449 R2132).

27. 'Comparative Agricultural Statistics for the Transvaal: 1908–10', p. 5.

28. The mines consumed about 45 per cent of all cattle slaughtered on the Rand, or nearly 20 per cent slaughtered nationally: *Official Year Book, no. 16, 1933–34* (Pretoria: 1935), p. 404; *Official Year Book, no. 18, 1937* (Pretoria: 1937), pp. 829–30; *Annual Report of the Department of Agriculture, for the Year Ended 30 June 1934*, reprinted in *Farming in South Africa*, December 1934, pp. 404, 488 (UG).

29. Magistrate Vryburg to Sec. for PM, Pretoria, 23 October 1922 (SASA LDB1449 R2132); N. Parsons, 'The Economic History of Khama's Country in Botswana, 1844–1930', in R. Palmer and N. Parsons (eds), *The Roots of Rural Poverty in Central and Southern Africa* (London: 1977), p. 131.

30. D. C. M. Platt, *Britain's Investment Overseas on the Eve of the First World War* (London: 1986), pp. 98–100; I. Phimister, 'Meat and Monopolies: Beef Cattle in Southern Rhodesia, 1890–1938', *Journal of African History*, 19/3 (1978), p. 400.

31. S. B. Woollatt, 'Stock Industry of South Africa, Part 2', *South African Journal of Industries*, 2/8 (1919), pp. 1058–9.

32. *Farmers' Weekly*, 4 September 1935, p. 1867.

33. *Annual Report of the Department of Agriculture, 1907–8*, pp. 57–8 (TG); 'The Cattle Industry', a Reply to Mjr. E. W. Hunt, President SAAU (n.d., ?1927) (SASA LDB1451 R2132(5)).

34. L. Bagshawe-Smith, 'Cattle Breeding', *Transvaal Agricultural Journal*, 8 (1910), pp. 215–16; Cowen and Shenton, *Doctrines of Development*, pp. 174, 184, 209–12.

35. G. Roper, BOT to Union HC, London, 16 July 1915 (SASA LDB629 R656(2)); Union Trade Commission, London, to Secretary of Agriculture, Pretoria, 23 July 1931 (LDB647 R656/50(2)).

36. 'Editorial Notes – An Export Trade in Meat?', *South African Agricultural Journal*, 7/6 (1914), p. 792.

37. *Report of, and Evidence to, the Native Land Committee – Western Transvaal (1918)*, Zeerust, 9 October 1917 (UG31-18).

38. Minutes of the Board of Directors, vol. 14, 1 February 1916 (Imperial Cold Storage [ICS], Barlow Rand Archives [BRA]); I. G. J. van den Bosch, 'Report on Cattle Industry in South

Africa', *Department of Agriculture: Local Series Bulletin*, 11 (1917), p. 6; *Agricultural Census*, 1921, table 7, p. 4 (UG44-22).

39. *Official Year Book, 1928–9*, p. 388; 'The South African Export Trade in Beef: Report on Recent Consignments', *Department of Agriculture, Farmer's Bulletin (Local Series)*, 83 (1919), p. 1.
40. 'Export of South Africa's Export Trade in Meat: Messrs. W. Weddel & Co's Review', *South African Journal of Industries*, 4/7 (1921), p. 530; Secretary of South Magaliesburg Farmers' Association to Minister for Agriculture, Cape Town, 22 March 1923 (SASA LDB1449 R2132(1)).
41. George Pott, Johannesburg, to Under Secretary for Agriculture Williams, Pretoria, 7 January 1924 (SASA LDB1449 R2132(2)).
42. *Farmers' Weekly*, 13 September 1939, p. 2055.
43. Dir. of Ag, Salisbury, to PVO Chase, Mafeking, 18 October 1923 (Botswana National Archives [BNA]: S18/4); Resident Commissioner, Bechuanaland Protectorate, Mafeking, to High Commissioner, Cape Town, 1 February 1935 (S275/4).
44. Secretary of South Magaliesburg Farmers' Association to Minister for Agriculture, Cape Town, 22 March 1923 (SASA: LDB1449 R2132(1)); *Farmers' Weekly*, 5 December 1928, p. 1269.
45. Director of Veterinary Services P. J. du Toit to Dist. Commdt. Rustenburg, 22 July 1937 (SASA: South African Police [SAP] 162 30/2/38).
46. *Farmers' Weekly*, 26 November 1930, p. 955.
47. Chief Magistrate Umtata to Secretary for Native Affairs, Pretoria, 13 January 1937 (SASA: Native Affairs Department [NTS] 7335 127/327 (2)).
48. Memo, Deputy DFAH Romyn to DFAH Thornton, 17 November 1926 (SASA LDB2046 R2835).
49. Chief Magistrate Umtata to Secretary for Native Affairs, Pretoria, 13 January 1937 (SASA NTS7335 127/327 (2)).
50. *Debates of the House of Assembly*, 18 May 1934.
51. *Farmers' Weekly*, 15 January 1919, p. 2084; 2 June 1926, p. 1176; 10 November 1926, p. 861; Secretary for Justice to Secretary of TAU, Pretoria, 14 February 1933 (SASA: Department of Justice [Jus] 2171/474/12).
52. Magistrate G. Rustenburg to Secretary for Agriculture, Pretoria, 27 March 1934 (SASA Jus606 1307/34).
53. *Annual Report of the Department of Agriculture, for the Year Ended 30 June 1939*, reprinted in *Farming in South Africa*, December 1939, p. 552.
54. *Farmers' Weekly*, 31 August 1938, p. 1872; 22 March 1939, p. 29.
55. *Farmers' Weekly*, 19 April 1939, p. 355.
56. *Farmers' Weekly*, 22 December 1937, p. 1024.
57. *Farmers' Weekly*, 15 November 1939, p. 677.
58. Union of South Africa, *Reconstruction of Agriculture: The Report of the Reconstruction Committee* (Pretoria: 1943), pp. 32, 40, 43.
59. Union SA, *Reconstruction of Agriculture*, pp. 33–4.
60. *Farmers' Weekly*, 24 December 1924, p. 1559.
61. D. Grossman, 'Relations between Ecological, Sociological and Economic Factors Affecting Ranching in the North Western Transvaal', PhD thesis (University of Witwatersrand: 1988), pp. 98–103.
62. Grossman, 'Relations', pp. 98–103.
63. B. Cousins, 'Livestock Production and Common Property Struggles in South Africa's Agrarian Reform', *Journal of Peasant Studies*, 23/2/3 (1996), p. 202.

PART 5

Comparing Settler Societies

Empire and the ecological apocalypse: the historiography of the imperial environment

John M. MacKenzie

The environmental history of the British and other European empires has been one of the great growth areas of contemporary historical scholarship. Historians of science, medicine and natural history, geographers and natural scientists have all contributed to this burgeoning field, creating in effect a completely new subdiscipline. More recently, cultural historians have also become active in the field. Nevertheless, notable American practitioners like Donald Worster and Alfred Crosby have re-emphasized both the alleged American origins and continued domination of environmental history.[1] This injection of nationalism is ironic, since Worster has argued that environmental history constitutes the major replacement for historians' concentration on the history of the nation-state in the late nineteenth and early twentieth centuries. In any case, as with all nationalist interpretations, there has been a rapid and spirited response. American pre-eminence has been contested by Richard Grove, both in *Green Imperialism* and in a recent paper.[2] At first sight this may seem a relatively sterile debate, but it has had the useful effect of uncovering the multidisciplinary sources of modern historical concerns.

Worster and Crosby have stressed the moral roots of environmental studies in the development of the (American) green movement of the early 1970s. They can equally be distinguished in the moral climate of decolonization and European concerns about the imperial interaction with the wider world. This notable strand of environmental history is now sufficiently broad for it to be possible to distinguish at least four historiographical tendencies within it, and new approaches are continually being uncovered, not least by the contributions to this volume. These four can be briefly characterized as the apocalyptic, the neo-Whiggish, the longer perspective, and the fully integrated cultural schools. They are not, of course, mutually exclusive and other defining modes can be overlaid upon them; for example, it is possible to distinguish Eurocentric, peripheralist, neo-centric, and ethnic perspectives. Moreover, some at least of these developments can be seen to mirror the intellectual and practical odyssey of imperial rule itself, from arrogant self-confidence to apprehensive questioning and doubt.

Empire, power and the apocalypse

Before examining the apocalyptic school of imperial environmental history, it is necessary to turn to imperialists' estimation of themselves, for, as frequently happens in the discipline of history, modern analyses often stand past ideologies on their heads, at least in identifying the gulf between objectives and results. It used to be thought that, if western European empires had anything positive to offer the rest of the world, it was surely their capacity to act as the bearers of the scientific, medical and engineering cargo upon which they ultimately based their claims of superiority. If, in the words of Michael Adas, 'machines' were 'the measure of men', then Europeans clearly perceived themselves as giants.[3] The kind of self-confidence offered by this sense of technical power comes through in David Livingstone's conviction in the positively redemptive powers of steam engines – even if they seldom worked for him as he hoped – in Rudyard Kipling's fascination with machines and the potential of the engineer to dominate and harness nature, and in the countless examples of contemporary wonder at the development of marine engineering, machine tooling, the submarine cable and the railways. Daniel Headrick has built a career out of arguing for their importance in his books *The Tools of Empire*, *The Tentacles of Progress*, and *The Hidden Weapon*.[4]

One of the objections to Headrick is that he takes Europeans too much at their own estimation. Certainly, their overweening environmental confidence, founded on such technical progress, can be found in any number of sources throughout the nineteenth and twentieth centuries. The Scots missionary Robert Moffat perceived environmental control as the distinctive characteristic of the Christian, contrasted with the heathen African's alleged helplessness. An engraving of his mission in his book *Missionary Labours and Scenes in Southern Africa*, published in 1842, reveals ordered hedges, paths, plantings and buildings contrasting with the wildness beyond.[5] Such polarities appeared in the illustrations to countless works on settlers and their power.[6] Livingstone's vision of great cotton fields down the Zambezi, populated by the poor of the central belt of Scotland, was surely influenced by the dramatic changes that he himself had observed in the agriculture of the Scottish Lowlands, in enclosure, draining, selective breeding, new approaches to hydrology and the rest.

For Sir Charles Eliot, Commissioner of the East Africa Protectorate at the turn of the nineteenth and twentieth centuries, the problem with Africa was precisely that its environment required to be controlled and transformed. The past of the continent was 'uneventful and gloomy' because of the lack of contact with the outside world as a result of the natural obstacles, deserts, marshes or jungles which separated the coast from the interior. He went on:

> Nations and races derive their characteristics largely from their surroundings, but on the other hand, man reclaims, disciplines and trains nature. The surface of Europe, Asia and north America has been submitted to this influence and discipline, but it has still to be applied to large parts of South America

and Africa. Marshes must be drained, forests skilfully thinned, rivers be taught to run in ordered course and not to afflict the land with drought or flood at their caprice; a way must be made across deserts and jungles, war must be waged against fevers and other diseases whose physical causes are now mostly known.

It is a fascinating statement. Having slid smartly from environmental determinism to ecological control, he applies the language of discipline and training to nature in the same way in which it was invariably used of indigenous peoples. Natural forces, like people, were to be acculturated to the modern world. Ronald Ross's final exposure of the causes of malaria had clearly convinced him that the caprice of the microbe could be ordered like that of the flood. In a final peroration, he asserted that 'this contest with the powers of Nature seems a nobler and more profitable struggle than the international quarrels which waste the brain and blood of Europe and Asia.'[7] Sir Charles Calwell's characterization of small colonial wars as 'campaigns against nature' becomes a battle with the environment itself instead of with other humans.[8]

This pride in environmental control was expressed in countless other ways. It can be found in the rolling periods of the purple prose of Viceroy Curzon's speeches at the opening of Indian bridges; in the two enormous recumbent lion statues that the British installed to guard the ends of the great Ganges Canal, imperial hydrological despotism expressed through the king of beasts; or in the creation everywhere of zoos, menageries and botanic gardens by imperial governors in their gubernatorial residences, a classic and symbolic taming of nature in the very backyards of the rulers of empire. It can also be found in the tremendous puffing of the resources of Africa by early explorers and commissioners like Sir Harry Johnston and Sir Arthur Hardinge.[9] This propaganda continued throughout the era of imperial rule. It was still being projected in the rapturous descriptions of such imperial environmental designs as the groundnut scheme in Tanganyika in 1947 – 'solid ground for hope, hundreds of miles of jungle cleared by science and the bulldozer with a real promise of a better life for African and European' – or in the movement of people and animals consequent upon the building of the Kariba Dam and the formation of the vast lake in the 1950s, a project which came to symbolize and even justify the very political unit of the Central African Federation.[10] Sir Harry Johnston portrayed the shift from assurance to anxiety in his own career and writings. He regarded himself as a natural history collector, zoologist and artist before he was an explorer and administrator.[11] He wrote ecstatically of the economic potential of Africa and its natural attributes, creating botanic gardens and small zoos wherever he established a government house. However, he also expressed mounting alarm at degradation and decline. Like so many natural history enthusiasts and hunters of the period, he was particularly anxious about the decline of animal numbers. When he had visited Tunisia in the late 1870s, he had found it still full of big game. When he returned as Consul-General in Tunis in 1897, the game had already disappeared. He joined the Society for the Preservation of the Wild Fauna of the Empire when it was founded in 1903 and was active in its demands for stricter controls upon African hunting.[12] Even more interestingly, he has an almost throw-away line about Tanganyika in his autobiography, published in 1923. When he had visited the Nyasa-Tanganyika plateau in what is now south-western Tanzania in 1890, he had seen excellent crops, a profusion of wild flowers and an abundance of game: 'The Tanganyika in those days was a paradise; later it was to be ravaged by wars, depopulated by sleeping sickness and afflicted in many other ways.'[13] In the rivalries of the partition, Johnston was a notable Germano-phobe and there can be little doubt that he was implicitly ascribing this degradation to German rule.

This is indeed a characteristic of the apprehensive imperialist: the agency for ecological decline was invariably placed elsewhere. This was true of the disappearance of African game and the decline of forests and increased desertification.[14] British foresters in India (probably more than the Germans, whom the British employed) worried constantly about the damage caused by indigenous forest dwellers.[15] When E. P. Stebbing visited West Africa in the 1930s, he attributed the alarming denudation of tree and bush cover to the damaging effects of African pastoralism and shifting cultivation.[16] Indeed, swidden agriculture and the use of fire were excoriated everywhere by imperial rulers and their technical advisers. In the interwar years, irrigation engineers in India could not fail to observe that the grand canal schemes of the British were going wrong, but they were all too ready to place the blame upon poor maintenance and misuse by the agriculturalists whom they were supposed to benefit. These observations form a ready bridge to the apocalyptic school of imperial environmental history.

The apocalypse

Elizabeth Whitcombe's pioneering work on the canal systems of British India, published as early as 1972, illustrates this beautifully.[17] She demonstrated that British engineers and agronomists set about the amazing canal developments of the twentieth century with an environmental zeal that can only be described as religious. The British set about rebuilding and massively extending the canal systems of the Yamuna (Jumna), Ganges and Indus Rivers in north India and the Cauvery and Godavery in the south. They had a complex of motivations: extended settlement would increase the land revenue, the fiscal basis of their power; they would yet again find a means of fitting themselves into the Mogul legacy; by overcoming intermittent precipitation and groundwater shortages, they would illustrate command of the environment. The results were, however, very different from those intended. Since both the system and its execution were misconceived, it produced not economic regeneration, but extensive and damaging waterlogging, as well as high levels of salination akin to those found in ancient Middle Eastern irrigation systems which had similarly gone wrong.

More recently, Whitcombe has written of the medical consequences in the resultant expansion in the incidence of malaria.[18]

Whitcombe's work has a magisterial coolness about it, belying the heat, dust and hydrological rush of its subject. Perhaps the prime early and hotter example of the historiography of the imperial apocalypse is to be found in the publications of Alfred Crosby. If, for Whitcombe, the grand environmental projects had gone wrong, Crosby saw Europeans as initiating a successful biological conquest of the globe. In both his *Columbian Exchange* of 1972 and his more ambitious *Ecological Imperialism* of 1986, he painted a picture of organisms of all sorts being marshalled, consciously and unconsciously, for just such a campaign.[19] Mammals, birds, freshwater fish, insects, pathogens, trees, plants and weeds set about the creation of neo-Europes, exotic environments comprehensively overlaid with the extensive biota of the new conquerors. These events were promoted by economics, aesthetics, sport, nostalgia, or simply absent-mindedness and inefficiency. Yet his vision was not entirely global, for Crosby paid little attention to Africa and he also argued that the well-established historic peasant cultures of Asia had been able to resist these processes, a contention that some modern Indian scholars deny.[20] What is more, Crosby suggested, highly dubiously, the surprising thing was that so little came back. In his determination to see biological imperialism as a one-way process, illustrated by the imperialist urges of the dandelion, he seemed to know little of the expansion of the eucalypt and Australian wattle, the depredations of the rhododendron, Japanese knotweed or Himalayan balsam, the territorial hunger of the grey squirrel, the mink or the New Zealand flatworm.

Meanwhile Lucile Brockway had already provided a conspiratorial twist for this biological expansion by seeing continental and intercontinental plant transfers as part of a global plot masterminded by scientific controllers at Kew Gardens.[21] Moreover, as many other environmental and economic historians have pointed out, rather more convincingly, such plants, in their frequent transformations from foraged to cultivated product, spread plantations throughout the world. And such plantations created maximum social and environmental damage through being land-extensive and soil- and labour-intensive. Vast tracts of pre-colonial nature were overwhelmed as sugar, coffee, tea, indigo, the opium poppy, cinchona, jute, sisal, tobacco and rubber marched across the landscape. These plants were the shock troops of economic and natural historical warfare.

Studies in East and Central Africa powerfully developed this sense of imperial catastrophe. Helge Kjekshus, in his *Ecology Control and Economic Development in East African History*, strongly contrasted images of a period of plenty in pre-colonial times with the shattering effect of a series of environmental and medical disasters attendant upon the arrival of Europeans in the 1890s.[22] Some of these, like rinderpest, afflicting both cattle and game, smallpox and jigger fleas, menaces to human health, were introduced directly, albeit inadvertently, by European agency. Others, like the prevalence of drought and the spreading of locust swarms, happened to coincide with the appearance of Europeans, leading contemporary Africans to draw appropriately hostile conclusions. Others again, like the spread of nagana and East Coast fever among cattle and sleeping sickness among humans, were the results of misconceived colonial policies. In rather more sophisticated studies spanning parts of Zambia, Malawi and Mozambique, Leroy Vail has argued that a 'major ecological catastrophe' resulted from the combined impact of expanding capitalism and colonial administration in the region.[23] If some evidence of pre-colonial problems can be identified, then imperial rule seized a system that was already under stress and pushed it over the edge.

To heighten this sense of an imperial apocalypse, historians and others have felt it necessary to offer a constrasting image of a pre-colonial past that was in harmony or balance with nature. Kjekshus has been criticized for creating just such a vision of 'Merrie Africa' – and, indeed, parallel images of 'Merrie Australia' and 'Merrie India' can be found in the literature. William Lines's *Taming the Great South Land* of 1991 is a record of rapine and plunder, of the piling of environmental disaster upon natural catastrophe since the arrival of Europeans in Australasia. Whereas, according to Lines, Aboriginal occupation had only touched the environment lightly and did 'not greatly disturb relationships within the community of plants and animals', Europeans brought destruction in their wake.[24] Such a view is hardly sustainable, as had been suggested by Bolton several years earlier.[25] In *This Fissured Land* (also of 1991), Madhar Gadgil and Ramachandra Guha create a theory of modes of resource use to illustrate the greater harmony between humans and nature in the pre-imperial period in India.[26] In this and other works by Guha, Indian hunters, pastoralists and cultivators are all seen as promoting sustainable yield policies as well as establishing mutually beneficial ecological niches. Europeans disrupted and destroyed these fine balances, not least in their exaggerated and exclusivist forest policies.

Such visions of global apocalypse have been assiduously fed through into populist green histories. Clive Ponting's *Green History of the World* presents a strikingly doom-laden picture.[27] In his reading, it is not only a case of 'Apocalypse Now', but also of 'Apocalypse Then'. Influenced by Marshall Sahlins's *Stone Age Economics*, Ponting, like some other popular writers, fingers the Neolithic revolution as the start of human ecological madness. Since then, successive civilizations have been doomed to destruction through self-inflicted environmental degradation. This dramatic counter-progressivism views world history as one long free fall, with imperialism as its global accelerator. The entire past is coloured with fear of the future.

Neo-Whiggism

It is perhaps inevitable that the post-modernist age should have rediscovered a powerful progressive antidote to this 'apocalyptism'. This tendency privileges European sensibilities in producing environmentalist ideas from the seventeenth, eighteenth or nineteenth centuries. It can be dubbed neo-Whiggish, because it does indeed chart progress through the development of the bourgeois intellect. The model is perfectly symbolized by the word 'roots', which tends to appear frequently in its titles.[28] It fits into long-standing Eurocentric and Anglocentric traditions, which have been developed particularly in the last sixty years or so. It can be found in the work of the sociologist Norbert Elias, *The Civilizing Process* of 1939,[29] which charted the development of manners in late eighteenth- and early nineteenth-century Europe, or again in the words of Harold Perkin, who suggested that: 'between 1780 and 1850 the English ceased to be one of the most aggressive, brutal, rowdy, outspoken, riotous, cruel and bloodthirsty nations in the world and became one of the most inhibited, polite, orderly, tenderminded, prudish and hypocritical'.[30] Famously, Keith Thomas carried this notion into the English – and his work is highly Anglocentric – relationship with nature. The science of the Enlightenment, as well as of the Romantic and post-Enlightenment periods, produced a 'revolution in perceptions' which created 'new sensitivities that have gained in intensity ever since'.[31] David Allen and Harriet Ritvo, both of them in well-contextualized works that give due attention to both class and power, tended to shift these growing sensitivities from the beginning towards the end of the nineteenth century.[32]

While James Serpell has identified the moral contradictions in the human approach to domestic and wild animals, the philosopher Mary Midgley has also analysed nineteenth-century hunting works in terms of heightened sensibilities.[33] She has even argued, wholly unconvincingly in my view, that the Highland butcher of a Nimrod, Roualeyn Gordon Cumming, clad in his kilt and Badenoch brogues, demonstrated 'a true belief in the consciousness, complexity and independence of the victim'. In suggesting that apparent cruelty towards elephants is not necessarily analagous to callousness towards people, Midgley demonstrates an inadequate understanding of the vast range of imperial hunting literature and of imperial campaigns, in which hunting imagery was applied to humans right down to the time of Mau Mau in Kenya in the 1950s.[34]

In some respects, though decidedly not in others, Richard Grove writes within this tradition, though his imperial focus and his command of primary material is greater than that of all his predecessors. In his defiantly titled *Green Imperialism*, he has been involved in identifying the roots of environmental ideas as lying much further back in history than has ever occurred to the American practitioners, blinkered as they are by the nationalist obsession with George Perkins Marsh, John Muir and Henry David Thoreau.[35] Through his study of ecological ideas relating to oceanic islands, the development of desiccation theory and anxieties about deforestation and species extinction, Grove has convincingly demonstrated not only the antiquity of such environmental thinking, but also its international and peripheral character. For him, the key ideas come not from the European metropolis, but from the periphery, and are relayed through international scientific networks, particularly those of the French and the Scots. By an attractively neat analytical sleight of hand, he has linked such ideas to radical politics in the late eighteenth century.

He has also provided a significant ethnic context to the development of such ideas, not only through the capacity of colonial ecologists to draw on indigenous knowledge, but also through the particular interests and expertise of the Scots. The botanist doctors of the Indian Medical Service, largely trained in Scottish universities, were the intellectual propagators of such ideas within India, the Cape and elsewhere in the British Empire.[36] Although Grove's work has considerable strengths (not least in its remarkable globalization) and offers strikingly new interpretations, it often privileges ideas over policy, almost suggesting that the former lead ineluctably to the latter. In any case, he gives hostage to fortune by ending *Green Imperialism* in 1860, just as the exploitative force of imperial rule moves up several gears with the working through of the 'second industrial revolution' of the period. It should be said, however, that other publications of Grove have noted the economic shifts and the constraints and barriers to environmental ideas in the political, social and cultural contexts of late nineteenth-century imperialism.

The longer-perspective school

As fresh historiographical schools continue to emerge, it is no longer possible to see this third strand as the final element in that satisfying rule of three that has so often been a central feature of philosophy, culture and the arts. It decidedly cannot be privileged within a challenge, response and resolution paradigm. Nevertheless, the fundamental problem with both 'apocalyptism' and 'neo-Whiggism' is that, in their different ways, they ascribe too much power to empire. The British Empire, vast and apparently despotic as it seemed, was in reality a ramshackle conglomerate, very far from the all-seeing, all-powerful monolith envisaged by Edward Said and his followers among the discourse theorists.[37] It was decentralized and highly heterogeneous, bearing within it many different types of rule as well as social, economic and racial systems. What is more, its influence was felt in distinct parts of the globe over very different time-spans.

Perhaps it is significant that this third school has been developed largely, though not exclusively, in the case of Africa, where the imperial period has been characterized (in one book title at least) as *The Colonial Moment in Africa*.[38] As the post-colonial era lengthens, perspectives and time-scales have tended to open up. Much new work, particularly in Africa, has reduced the tendency to see the imperial experience as both profoundly transformatory and uniquely destructive. A great deal of this new work has been concerned with fragile ecologies, with forest and marginal zones, with regions of transhumant pastoralism, with faunal extinctions and survivals, with issues involving relationships between peoples and power, demographic and climatic change and the incidence of famine.

Much of this research has tended to see the changes wrought by imperial power as but one phase in much longer cycles of environmental ups and downs not unlike those of the 'dismal science' of economics. Indeed, indigenous knowledge in many regions of Asia, Africa and Australasia reveals that many peoples have their own awareness of some form of the biblical cycle of feast and famine. At the same time, climatic history has been catching up with its sophisticated use in the natural sciences, and historians and archaeologists are increasingly coming to grips with pluvials and inter-pluvials, little ice ages, volcanic and El Niño-induced transformations. Linguists, historians and anthropologists have revealed words for 'dearth', like that powerfully expressive word of the Shona of Zimbabwe, *shangwa*.

Moreover, pre-colonial peoples had more power to transform their environments, mainly through fire, than imperial rulers or modern scholars have ever allowed. This is true, as we now know, of Australia, India and Africa. In comparatively recent times, there were almost certainly pre-colonial species' extinctions caused by overhunting and, at times, profligate killing. Examples of the latter have been found in North America and Australia.[39] The arrival of new migrant peoples, like the Bantu-speakers in Africa or dominant élites in India, had the capacity to transform the human relationship with botanical and zoological contexts as much as, or, in some cases, more than, colonial rulers, not least because they had a longer time to do so. Hunters and gatherers were perhaps well aware of this: there is a celebrated bushman cave-painting not far from Harare in Zimbabwe which, very movingly, depicts an immigrant Bantu-speaker cutting down a tree with an axe, an action which must have been technically and environmentally inconceivable to the painter.

At any rate, the repeated incidence of dearth must have produced both human and zoological demographic swings. In the African case, Europeans almost certainly arrived during one of a long series of environmental downturns, which both indigenous contemporaries and modern protagonists of the apocalyptic view attributed to their agency. Thus we have to understand the mutual effects and complex oscillations of both the natural cycle and human-induced change. We now know more of the historical depth of famine in, for example, both Ethiopia and India, knowledge which in both cases goes back to the sixteenth century and earlier.[40] We know that deforestation is far from being just a modern phenomenon; nor is the tight control of forests, their resources and who may live within them. Recent research has indicated the scale of environmental degradation in Indian forests under the Moguls, as well as the manner in which successor states to the Mogul Empire may have developed forest policies which became a model for the British at a later date.[41] There has been a good deal of speculation about the extent to which ecological problems had effects not only on a medieval state like Zimbabwe, but also on eighteenth- and nineteenth-century African polities in Zululand, Angola and Malawi.

Other scholars have pointed to the complex diversity of the imperial impact. McCracken, for example, has suggested that capitalism, in the shape of commercial tobacco-growing in Malawi, interacted with environments rather than dominated them, producing a mix of deleterious and favourable outcomes.[42] In any case, environmental enlightenment is not the sole prerogative of any one side in the imperial relationship. At times, indigenous peoples succeeded in frustrating attempts at botanical and forestry protection – examples have been found in both West and East Africa.[43] Moreover, the imperial monolith has increasingly fragmented. Experts and administrators sometimes tried desperately to settle nomadic pastoralists, not always successfully; elsewhere, pastoralists were culturally valued more highly than the supposedly softer, stationary peasantry. Some colonial authorities in Africa sought to destroy game to try to beat back the incidence of the tsetse fly, which used game as a host; others created vast national parks to encourage the regeneration of game stocks.[44] The policy pursued largely depended on whether the territory contained white settlers with cattle to be protected. As always, expert opinion was highly attuned to the political contexts that it served.

Towards the end of imperial rule, there were at least the beginnings of a better understanding of the interrelationship between forest peoples and their environment and between pastoralists, their herds and game. The nationalist historiography has often influenced historians of natural history to concentrate on instances of resistance to European policies, when submission and collaboration may have been just as prominent a part of indigenous responses. In many cases, post-colonial states have been more susceptible to sectional interests than imperial rulers.[45] No modern state likes people to move around, and many post-colonial states have been even more concerned to settle pastoral nomads than their colonial predecessors. Hunters and gatherers invariably come in for a raw deal, as recent examples in Africa and elsewhere demonstrate repeatedly. What is more, such states have often proved more responsive to powerful international conservation lobbies which do not always take indigenous needs into account. Just as the longer perspective school can dip deeper into the past, so, too, can it come closer to the present.

The fully integrated cultural school

This tendency in environmental history is a distinctively modern one, insofar as it often deals with constructions of nature as much as the supposed realities. It also attempts to set environmental issues into their full economic, political and cultural contexts. In the process, however, it has often tended to re-nationalize environmental history. In the past, the great strength of environmental history has been its capacity to transcend national, regional or even continental boundaries. This has certainly been the case with the work of Crosby, Grove and others. In *The Empire of Nature*, I very self-consciously wrote about both Africa and India in an attempt to demonstrate aspects of the common scientific, cultural and legal cultures that obtained throughout the British Empire.[46] In that work, I also set out to place imperial hunting into both indigenous and metropolitan cultural contexts. I argued that conservationist policies had to be understood not in terms of the development of sensitivities, but as ideas that were only possible once the economic need for the exploitation of animals had begun to pass away. They also had to be analysed – together with the legislation that they spawned – within their racial, scientific and settler environments. As alarm about the decline of animals increased in the 1890s and early years of this century, European hunters produced an apocalyptic vision which often produced equally apocalyptic solutions: the creation of vast reserves and national parks, the movement of peoples, widespread culling of both domestic and wild animals, particularly so-called 'vermin', and the imposition of hunting bans that were highly culturally determined. Ultimately, many of these policies were as disastrous, to the interests of both humans and animals, as the problems they were designed to overcome. This was particularly the case with the spread of tsetse fly and the incidence of nagana and sleeping sickness.[47]

This kind of cultural approach has been developed in much more sophisticated ways in recent times. *The Kruger National Park* by Jane Carruthers has an importance far beyond its relatively brief length or apparently specialist focus.[48] Her subtitle, 'A Social and Political History', could perhaps be expressed more accurately as a cultural and racial history. She studies the development of that vast park not only in terms of the lives of Africans, Europeans and animals interacting with each other through hunting, subsistence, war and leisure, but also in the context of Afrikaner nationalism. Although Afrikaners often paid no more than lip-service to conservationist measures, they soon recognized the significance of the Kruger Park not only for their wilderness myths, but also for their search for international acceptability, particularly once the full nationalist racial programme had been inaugurated after 1948. The nakedness of apartheid was clothed in the fig-leaf of the conservationist Kruger.

Tom Griffiths's superbly suggestive *Hunters and Collectors: The Antiquarian Imagination in Australia*, approaches constructions of the environment through successive interpretations of the human past.[49] He analyses the controversies about so-called wildernesses and the preservation of ecologies complete with their palimpsests of human endeavour superimposed or interleaved within them. This rich blend includes issues of tourism, the often contradictory phases of ecological management and preservation of the built environment, as well as private and museum collecting and their related exhibitions. Hunting and collecting took place within a landscape that was repeatedly being re-evaluated by settlers, even as their relationship with Aborigines, geological and human time-scales, and their own ancestors was progressively transformed.

Geoff Park's *Ñga Uruora* ('the groves of life' in the Maori language) brings together a personal and romantic experience of landscape with a sustained analysis of the Maori and *Pakeha* (white) approaches to exploitation, degradation and sustainability in the fertile coastal plains of both North and South Islands of New Zealand.[50] It also explores the responses of art and photography to these lands, where survival, economics and spirituality profoundly intermingle. There is, perhaps, a tendency towards a pre-colonial 'Merrie New Zealand' here, and the repeated interposition of the author's own personal responses is reminiscent of Simon Schama's *Landscape and Memory*.[51] Schama, however, renders his study of the nationalist constructions of landscape within Europe almost unreadable through his labyrinthine, post-modernist and obtrusively personal approach.

The partial re-nationalization of environmental history by Carruthers, Griffiths, Park and Schama is not necessarily a bad thing. Constructions of nature inevitably have a national or racial component. Additionally, these approaches represent the multilayered richness of the field and also offer all kinds of comparative methodologies useful elsewhere. They should help to promote, rather than hinder, the globalizing of environmental history. Indeed, Mahesh Rangarajan has recently asserted that the distinctive and extensive character of environmental studies in South Asia calls for a two-way process of global understanding and mutual fertilization.[52]

Other examples of the 'longer perspective' and 'culturally integrated' schools appear within this volume. Moreover, new neo-centric and peripheralist analyses can also be identified here. As the human past in Australia, as well as the antiquity of all its life-forms, is pushed further back, geographical as well as chronological perspectives can shift strikingly. A new prospectus repeatedly asserts itself, one which must develop indigenous conceptualizations of the environment, together with ethno-botany, ethno-entomology and natural history, and the capacity of Europeans to learn from these. The manipulation of the environment in the processes of resistance and collaboration must also be on the agenda, together with distinctive religious, philosophical and intellectual inputs. Since the histories of science and medicine have ceased to be the rather specialist and esoteric fields that they once were, there is also a need to develop the very productive work in these fields along with all the other cultural and ecological work in progress. It is abundantly apparent that four schools of environmental history represent no more than an opening bid.

Notes

1. Donald Worster, 'Doing Environmental History', in *The Ends of the Earth: Perspectives on Modern Environmental History* (Cambridge: 1988), pp. 289–307; Alfred Crosby, 'The Past and Present of Environmental History', *American Historical Review*, 100/4 (1995), pp. 1177–89.
2. Richard Grove, *Green Imperialism: Colonial Expansion, Tropical Island Edens and the Origins of Environmentalism, 1600–1860* (Cambridge: 1995); 'North American Innovation or Imperial Legacy? Contesting and Re-assessing the Roots and Agendas of Environmental History 1860–1996', paper delivered at the Australian National University Colloquium on the Environment, February 1996.
3. Michael Adas, *Machines as the Measure of Men: Science, Technology and Ideologies of Western Dominance* (Ithaca: 1989).
4. Daniel R. Headrick, *The Tools of Empire: Technology and European Imperialism in the Nineteenth Century* (Oxford: 1981), *The Tentacles of Progress: Technology Transfer in the Age of Imperialism, 1850–1940* (Oxford: 1988), *The Hidden Weapon: Telecommunications and International Politics, 1851–1945* (Oxford: 1991).
5. Robert Moffat, *Missionary Labours and Scenes in Southern Africa* (London: 1842), p. 147.
6. See, for example, Denis Cosgrove and Stephen Daniels, *The Iconography of Landscape* (Cambridge: 1988), p. 165.

7. Sir Charles Eliot, *The East Africa Protectorate* (London: 1905), pp. 4–5.

8. C. E. Callwell, *Small Wars: Their Principles and Practice* (London: 1906).

9. See, for example, Sir Arthur Hardinge, *Report on the Condition and Progress of the East Africa Protectorate from its Establishment to the 20th July 1897*, C 8683 (1897), p. 48.

10. Quoted in John M. MacKenzie, ' "In Touch with the Infinite", the BBC and the Empire 1923–53', in *Imperialism and Popular Culture* (Manchester: 1986), p. 183.

11. Sir Harry Johnston, *The Story of my Life* (London: 1923).

12. Johnston, *Story of my Life*, pp. 65–8; John M. MacKenzie, *The Empire of Nature: Hunting, Conservation and British Imperialism* (Manchester: 1988), pp. 211–16.

13. Johnston, *Story of my Life*, p. 276.

14. See, for example, David Livingstone, *Missionary Travels in South Africa* (London: 1857), p. 152. The question of indigenous agency in game destruction is explored in MacKenzie, *Empire of Nature*.

15. Madhav Gadgil and Ramachandra Guha, *This Fissured Land: An Ecological History of India* (Oxford: 1992); Richard P. Tucker, 'The Depletion of India's Forests under British Imperialism: Planters, Foresters, and Peasants in Assam and Kerala', in Worster (ed.), *Ends of the Earth*, pp. 118–40; Ramachandra Guha, 'Prehistory of Indian Environmentalism: Intellectual Traditions', *Economic and Political Weekly*, 4–11 January (1992), pp. 57–64; Jacques Pouchepadass, 'British Attitudes towards Shifting Cultivation in Colonial South India: A Case Study of South Canara District 1800–1920', in David Arnold and Ramachandra Guha (eds), *Nature, Culture, Imperialism: Essays in Indian Environmental History* (Delhi: 1995), pp. 123–51.

16. E. P. Stebbing, *The Forests of West Africa and the Sahara* (Edinburgh: 1937). Stebbing held (p. 5) that forest degradation occurred in three phases: 'farm it, graze it, then hunt it'.

17. Elizabeth Whitcombe, *Agrarian Conditions in North India* (Berkeley: 1972).

18. Elizabeth Whitcombe, 'The Environmental Costs of Irrigation in British India: Waterlogging, Salinity and Malaria', in Arnold and Guha (eds), *Nature, Culture, Imperialism*, pp. 237–59.

19. Alfred W. Crosby, *The Columbian Exchange: Biological and Cultural Consequences of 1492* (Westport: 1972), *Ecological Imperialism: The Biological Expansion of Europe 900–1900* (Cambridge: 1986).

20. Gadgil and Guha, *Fissured Land*, pp. 117-18.

21. Lucile H. Brockway, *Science and Colonial Expansion: The Role of the British Royal Botanical Gardens* (New York: 1979).

22. Helge Kjekshus, *Ecology, Control and Economic Development in East African History* (London: 1977).

23. Leroy Vail, 'Ecology and History: The Example of Eastern Zambia', *Journal of Southern African Studies*, 3 (1977), pp. 129–55, 'The Political Economy of East-Central Africa', in D. Birmingham and P. M. Martin (eds), *History of Central Africa*, vol. 2 (London: 1983), pp. 200–50.

24. William Lines, *Taming the Great South Land: A History of the Conquest of Nature in Australia* (Sydney: 1991), p. 11.

25. Geoffrey Bolton, *Spoils and Spoilers: Australians Make their Environment, 1788–1980* (Sydney: 1981).

26. Gadgil and Guha, *Fissured Land*, part 1.

27. Clive Ponting, *A Green History of the World* (Harmondsworth: 1991). For more respectable surveys, see I. G. Simmons, *Environmental History: A Concise Introduction* (Oxford: 1993); A. M. Mannion, *Global Environmental Change: A Natural and Cultural Environmental History* (London: 1991).

28. T. C. Smout, 'The Highlands and the Roots of Green Consciousness, 1750–1990', the Raleigh Lecture on History, *Proceedings of the British Academy*, 76 (1991), pp. 237–63; David Pepper, *The Roots of Modern Environmentalism* (London: 1984).

29. Norbert Elias, *The Civilising Process: The History of Manners*, trans. Edmund Jephcott (1939; Oxford: 1978).

30. Harold J. Perkin, *The Origins of Modern English Society* (London: 1969), p. 280.

31. Keith Thomas, *Man and the Natural World: Changing Attitudes in England, 1500–1800* (London: 1983), pp. 15, 243.

32. David Elliston Allen, *The Naturalist in Britain: A Social History* (London: 1976); Harriet Ritvo, *The Animal Estate: The English and Other Creatures in the Victorian Age* (1986; Harmondsworth: 1990).

33. James Serpell, *In the Company of Animals: A Study of Human–Animal Relationships* (Oxford: 1986); Mary Midgley, *Animals and why they Matter* (Athens, GA: 1984).

34. Midgley, *Animals*, pp. 14–16; Susan L. Carruthers, *Winning Hearts and Minds: British Governments, the Media and Colonial Counter-Insurgency 1944–1960* (London: 1995), p. 152.

35. Grove, *Green Imperialism*, pp. 2–3.

36. See Chapter 9, above; Grove, *Green Imperialism*, chapter 8.

37. Edward Said, *Orientalism* (Harmondsworth: 1985). For a critique of the post-colonial theorists, see John M. MacKenzie, *Orientalism: History, Theory and the Arts* (Manchester: 1995).

38. Andrew Roberts, *The Colonial Moment in Africa* (Cambridge: 1990).

39. Bolton, *Spoils and Spoilers*, pp. 6–7; Gadgil and Guha, *Fissured Land*, p. 73; William Beinart and Peter Coates, *Environment and History: The Taming of Nature in the USA and South Africa* (London: 1995), pp. 4–20.

40. See, for example, R. Pankhurst, *The History of Famine and Epidemics in Ethiopia prior to the Twentieth Century* (Addis Ababa: 1985); Vinita Damodaran, 'Famine in a Forest Tract', *Environment and History*, 1/2 (1995), pp. 129–58.

41. Michael Mann, 'Ecological Change in North India: Deforestation and Agrarian Distress in the Ganga-Jamna Doab 1800–1850', *Environment and History*, 1/2 (1995), pp. 201–20; Chetan Singh, 'Forests, Pastoralists and Agrarian Society in Mughal India', and Atluri Murali, 'Whose Trees? Forest Practices and Local Communities in Andhra, 1600–1922', both in Arnold and Guha (eds), *Nature, Culture, Imperialism*, pp. 21–48 and 86–122.

42. John McCracken, 'Colonialism, Capitalism and Ecological Crisis in Malawi: A Reassessment', in David Anderson and Richard Grove (eds), *Conservation in Africa: People, Policies and Practice* (Cambridge: 1987), pp. 63–77.

43. Andrew Millington, 'Environmental Degradation, Soil Practices and Agricultural Policies in Sierra Leone, 1895–1984', and David Anderson, 'Managing the Forest: The Conservation History of Lembus, Kenya, 1904–1963', both in Anderson and Grove (eds), *Conservation in Africa*, pp. 229–48 and 249–68.

44. MacKenzie, *Empire of Nature*, chapters 9 and 10. See also William Beinart, 'Empire, Hunting and Ecological Change in Southern and Central Africa', *Past and Present*, 128 (1990), pp. 162–86. An entire issue of the *Journal of Southern African Studies*, 15/2 (1989), was devoted to this and related issues.

45. Olusegun Areola, 'The Political Reality of Conservation in Nigeria', in Anderson and Grove (eds), *Conservation in Africa*, pp. 277–92; Madhav Gadgil and Ramachandra Guha, *Ecology and Equity: The Use and Abuse of Nature in Contemporary India* (London: 1995).

46. MacKenzie, *Empire of Nature*.

47. John M. MacKenzie, 'Experts and Amateurs: Tsetse, Nagana and Sleeping Sickness in East and Central Africa', in *Imperialism and the Natural World* (Manchester: 1990), pp. 187–212.

48. Jane Carruthers, *The Kruger National Park: A Social and Political History* (Pietermaritzburg: 1995).

49. Tom Griffiths, *Hunters and Collectors: The Antiquarian Imagination in Australia* (Melbourne: 1996).

50. Geoff Park, *Ñga Uruora: Ecology and History in a New Zealand Landscape* (Wellington: 1995).

51. Simon Schama, *Landscape and Memory* (London: 1995).

52. Mahesh Rangarajan, 'Environmental Histories of South Asia', *Environment and History*, 2/2 (1996), pp. 129–43.

CHAPTER 15

Empires and ecologies: reflections on environmental history

David Lowenthal

The diverse insights that unfold in this book are, at times, mutually at odds. Given the wide range of places, peoples and periods treated, divergent viewpoints are inevitable – and enlightening. We understand nineteenth-century New South Wales better by seeing how its settlement differed from, say, seventeenth-century Virginia. Six interrelated themes, converging on issues of environmental perception and impact, stand out as especially salient in the preceding chapters. I seek here to set them within a broader scope.

Imperial and settler types and stereotypes

Compelling images emerge out of Old World discovery, conquest and settlement abroad – images of rapture and terror in viewing exotic lands, images of fruition and rapine in engrossing them. But common stereotypes conceal profound differences of New World locale, of imperial ambition, of settler behaviour and of indigenous response. These variables are ecologically vital. To appreciate their import demands in-depth specific histories. We cannot understand the general ecology of empire without chronicling its particulars throughout Australasia and America, Africa and the Pacific. To know why imperial China gutted its woodlands while imperial Japan reafforested them, as Michael Williams notes, we need to study shifting imperial motives in both countries.[1]

The discussions that gave birth to this book focused predominantly on British imperial impacts. Vast extent and wide dispersal gave British imperial officials and scientists unrivalled opportunities to exchange environmental data and insights.[2] But collaboration was more than British; it was global. In learning about and modifying new worlds, European venturers and settlers moved with casual ease across imperial boundaries, much as the British in India imported French and German forestry expertise.[3] Caribbean history exemplifies such commingling. French and English seventeenth-century entrepreneurs hijacked Portuguese sugar-cane technology from Brazil to the West Indies, and deployed Dutch drainage and irrigation expertise to impolder the Guiana coast. Three centuries later, the quadri-imperial Caribbean Commission (established in 1946) sponsored programmes that ranged from area-wide climatic and volcanic data-gathering to the micro-level MABES scheme for five ecologically similar French, Dutch, and British islets (St Martin-Maarten, Anguilla, St Barthélemy, Sint Eustatius, Saba).[4]

In certain ways, however, British imperial ventures seem unique. Unlike Spain and, by and large, France, Britain made little use of indigenous labour – though, as Elinor Melville suggests, they might have used natives more had they had more of them.[5] In fact, finding but few and obdurate Native Americans, British and American planters relied on slave labour from Africa. It is ironic that the missionary priest Bartolomé de Las Casas's horrific exposé of Amerindian slaughter in Spanish America led to massive imports of African slaves, mainly into British and French colonies.[6] Yet British imperial enterprise, more than any other, depended on *settler* labour – English, Welsh, Scottish, and Irish. Reliance on settlers resulted in wholesale territorial engrossment. Alike in North America, Australia, and South Africa, settlement habitually advanced by occupying *all* the land, however unpromising initially. French and Spanish imperial impetus was far more centralized. Jesuit missions apart, France and Spain concentrated on areas of high economic reward, bypassing or abandoning others – such as, for Spain, most of the Caribbean islands. Economic viability in the Spanish imperium, unlike the British until they took India, required abundant *in situ* labour.

Yet British imperialists, like others, saw their colonies first and foremost as locales for extracting commodities to be exported. Local livelihood and ecology, indigenous or settler, were of no moment in themselves; all that mattered was producing as much as possible as cheaply as possible for the home market. 'The laws of commerce', as Edmund Burke wrote, 'are the laws of nature.' Emphasis on such aims outlasted formal decolonization. Even when merged with the University of the West Indies in the 1960s, Trinidad's Imperial College of Tropical Agriculture continued to concentrate on export commodities – sugar, cotton, coffee, cacao, bananas. Peasant crops and livestock were all but ignored, despite dawning awareness that they were crucial to the health of Caribbean environments.[7] Sugar, not yams, enriched European entrepreneurs. Indeed, when ex-colonial exports became unprofitable or politically problematic, the withdrawal of European interests often spelled catastrophe. Land management, no less than local employment, depended on plantation regimes. Abandonment of the Guyanese sugar industry has ruined the elaborate coastal drainage and irrigation system, leaving former estate villages at the mercy of flooding and resurgent malaria.

Lands long given up to imperial enterprise suffer by imperial withdrawal for two related reasons. Unless their economies remain neo-colonial, they are ill-equipped to bear the loss of habitual modes of employment. Nor can they rapidly overcome habits of dependence ingrained by long subservience to the will, as to the interest, of others. The plight of ex-imperial tropical Africa today exemplifies these problems.

Frontier and periphery

The dynamics of occupying and settling new worlds invariably raises questions about frontiers and peripheries. Notions of frontier as dream or nightmare, as challenge or obstacle, as imperial goal and as folk memory suffuse every settler society. What lies beyond the frontier is traditionally seen as 'empty' – that is, occupied only by indigenes. How frontiers are viewed shapes how people view themselves and how they treat their environment. Frederick Jackson Turner's concept of the frontier as the prime determinant of American character is unique to that country; South African and other contrasts repay further analysis.

Whereas Americans were schooled to view frontiers as fields of opportunity, the Australian frontier was more apt to be felt as a demeaning fringe – less a vital centre to be conquered than an impoverished periphery to be ignored. Those near the imperial edge came to see themselves, like their environments, as unimportant. To redress this self-demeaning imbalance, Tom Griffiths strives to show the centrality of the periphery.[8] Australia endured acute awareness of being peripheral in a double sense. One is its antipodal remoteness from the home country, being literally at the far end of the Old (the important) World. The other is its hollow centre, its dead heart, leaving Australia only peripherally alive. Had the centre any culture, it was only alien, Aboriginal. Some felt this flaw so fatal that it precluded ever achieving an Australian nation. No longer. The Aboriginality of Alice Springs is, for all but a few Australians, an emblem of union, not an omen of fracture.[9] Peripherality today is more psychic memory than actual menace. Phone and plane overcome or abate the drawbacks of the hollow centre. Nor is Australian demography uniquely peripheral: Brazilians lack a living centre; Scandinavia, Russia, and Chile are highly off-centred.

Most comparable to Australia is the United States of America. The myth of a Great American Desert long divided the continent; California was another country.[10] And the peopling of the Great Plains may come to be seen as only a brief interlude in its history. Rural demise today empties America's heartland, heralding fewer folk in Kansas than in Kakadu; Nevada, if not Nebraska, seems an American Nullarbor. The whole inter-montane West is seriously proposed as a Buffalo Commons divested of human habitation.[11]

Concern with peripheries seems peculiarly Anglo-Celtic. As noted above, the Spanish sloughed off their marginal colonies. Hispanic habit, perhaps? Old Castile similarly failed to integrate peripheral Iberian conquests; Basque, Catalan, and Andalucian adjuncts to this day remain, in patriotic and practical terms, beyond the Madrileño pale.

Ecological discourse: environmental determinism, human agency

Contrasting imperial British conquest with inexorable natural selection, Darwin's supporter T. H. Huxley thought 'the question of questions for mankind – the problem which underlies all others' – was to ascertain 'the place which

Man occupies in nature ... What are the limits of our power over nature, and of nature's power over us?' [12] Most pundits then believed that nature called the tune. Human acts, intended or accidental, could but dent the geological and biological fundament. 'Environmentalists' thought history and culture mainly determined by climate, land-forms, soils and vegetation. Mankind might improve nature but could not substantially transform it. [13]

Almighty nature is now dethroned. 'Environmentalism' today holds a reverse meaning: it expresses fears for a fragile nature degraded by human subversion of ecological order. Ecology became a recognized scholarly discipline only in the 1920s; but ecological awareness of human impact goes back to eighteenth- and nineteenth-century observers. Almost coincident with Huxley's question came an answer: George P. Marsh's *Man and Nature; or, Physical Geography as Modified by Human Action*. [14]

Environmental despoliation in imperial and ex-imperial realms – America, Australia, India – catalysed radical reassessment of nature, now seen as easily damaged by technology. Marsh's insights were notably imperial: he compared ongoing change in North America and alpine Europe with anciently devastated Roman, Ottoman and Hapsburg Empires. *Man and Nature* was persuasive to imperial stewards because it drew on modern forestry and engineering to assess environmental damage most evident in colonial and ex-colonial realms. Here deforestation, erosion and species extirpation went on at a fearsome pace. Men in charge of colonial lands faced agonizing choices between market-driven extractive enterprise and precautionary stewardship. More often than is commonly realized, they aimed, though usually in vain, to stem present-minded gutting of resources that, in Tom Dunlap's phrase, made rich parents – but poor children.

We inherit colonial habits along with degraded habitats. Again and again, environmental stewards seek to curtail entrepreneurial practices harmful to soils, vegetation, wildlife, even climate. Again and again, such efforts prove nugatory in the face of public ignorance, greed or unconcern. Again and again, catastrophe generates public demands for protection and renovation, followed by a new cycle of oblivion and ruthless exploitation.

Modern awareness of risks and hazards, along with increased concern about pollution, depletion, radiation, extinction, global warming and other threats, to some herald a post-imperial global reform. This seems to me overly sanguine. Public support for environmental causes is highly volatile and intensely parochial, as the Green movement's fluctuating fortunes show. Lip-service to such causes rarely translates into self-denying pragmatic commitment. Global prolegomena at Stockholm and Rio have yet to bind sovereign states in effective environmental collaboration.

Perceptions of settler impact

Settlers accounted for environmental change, whether desired or dismaying, as the work of God, of nature, of other people, or even of themselves. [15] In so doing they resembled those who stayed at home. But the extent and speed with which newly colonized lands were transformed excited the most impassioned praise or blame of some entity, divine or human, for fancied environmental weal or woe.

At the outset, settlers acclaimed their own impacts as essential and benign. Their prescribed mission – divine in the seventeenth century, civic in the eighteenth, imperial in the nineteenth – was to subdue and civilize the wilderness, along with its indigenous inhabitants. Over time, most saw themselves rather than a remote deity, as prime movers in this fructifying task.[16] An anecdote ascribed to the playwright Moss Hart, who in the 1930s converted a scruffy patch of New England into a fine country estate, epitomized this view. A visiting preacher enthused: 'What a beautiful place you have built here, you and God together!' 'Yes', Hart replied, 'and you should have seen it when God had it all to Himself.'

Early settlers saw nothing amiss with their conquest of nature. Untoward side-effects were small and easily repaired. Yet so profound was their cumulative impact, that unwanted alterations were soon too glaring to escape notice. In Australia and New Zealand as in North America, in South Africa as in the Caribbean, the ill effects of systematic logging, large-scale hunting and trapping, and the replacement of native species by imported plants and animals rapidly became patent. At first few settlers cared; shrugging off the damage as inconsequential or evanescent, they moved on to conquer new frontier environments. Not until the mid-twentieth century, when cumulative injuries threatened to wipe out environmental gains, did settler heirs take their forebears to task for their damaged legacy.[17]

Today's revulsion against damage and loss is as intense as past paeans of praise to nature's subjugation. Once despised wilderness is now adored. New World concerns are expressed most fervently – no such pioneering zeal drove Europeans who stayed put to conquer nature or to subdue indigenes, nor did any such remorse for past rapine later impel them to restore pristine nature.

Settler heirs now condemn their pioneering forebears as environmental destroyers – folk who, in Stephen Pyne's phrase, found a garden and left a wilderness. Efforts to restore the previous wild garden are now legion, as Jane Carruthers relates for South Africa.[18] In America, too, the National Park Service until recently took untouched nature for granted as the ideal. Scenic splendours were reserved for contemplative worship; every human impress had to be eradicated as a disturbing stain of 'progress'.[19]

Perceptions of indigenous impact

New World settlers and heirs habitually contrast their own environmental impress with that of native indigenes, so unlike themselves in culture and technology. (Old World peoples draw no such contrast, for they are their own indigenes.) Settlers generally judge indigenous impacts as slight compared with their own, but time has reversed the conclusions once drawn from this difference.

At the outset, imperial settlers were hardly aware of indigenous impacts, blind to signs of non-European occupation. They assumed that they saw virtually untouched virgin lands, 'almost fresh from the Maker's hands'. That indigenes without permanent farms or advanced tools had, over millennia, profoundly altered New World landscapes, and were still doing so, long went unrecognized. To be sure, it suited colonial incomers to overlook signs of native alteration; the apparent absence of indigenous 'improvements' helped to justify the removal of indigenes from tribal lands.

Indigenous impress seemed slight because European farmers and town-dwellers were ill-equipped to gauge the environmental effects of semi-nomadic hunters, gatherers, and shifting cultivators – above all, of the cumulative edaphic and floristic outcomes of fires periodically set. Any impacts that settlers did note seemed to them trivial, wasteful or unproductive. Indigenes unable or unwilling to abandon 'primitive' practices for permanent settlement were thus held doomed to give way to superior races with advanced technologies.

Like our views of settler impact, perceptions of indigenous impact have lately been turned upside down. Native influence is still seen as light, but lightness has become a virtue instead of a vice. Indigenes whose impact on nature was once dismissed as derisory are now stewards blessed with inherent environmental wisdom. The same mind-set that replaces confidence in technology with malaise about its fearsome effects exalts tribal peoples as ecological gurus. Aboriginal and Native American reverence of nature is held necessary to repair the ravages of technocratic greed.[20]

Just as indigenes changed their environments more radically than settlers realized, so too they not infrequently wrought environmental havoc. No culture has a monopoly on ecological sanctity. Indigenous ignorance or short-sightedness induced the extinction of moas in New Zealand, soil exhaustion in central Mexico, salinization in the Tigris/Euphrates, and wholesale 'eco-cide' among Arizona Anazasi and Guatemala Mayans.[21] To view indigenes as incapable of harm is as dehumanizing as earlier notions that they could do no good. Romantic primitivism jeopardizes realistic rapport with the environment.

Ecological purity and environmental chauvinism

In dislodging imperial doctrines, we risk succumbing to no less narrowing nativist dogmas. Regressive environmental and racial determinisms underlie the new mystique of the indigene as ecologist. Born close to nature, tribal folk are said to inherit talents for understanding and living with it. And since ancient occupation confers inborn environmental nous, indigenes become reputed as every land's most suitable occupants.[22]

Or even, in impoverished soils – as far as plants and animals are concerned – their *only* occupants. Although they accept multiculturalism, Eric Rolls and

Tim Flannery depict Australia as a paradigm of ecological fragility. So tenuous is the balance that the introduction of any alien plants, alien animals, alien agricultural practices is apt to devastate endemic species and may endanger the entire ecosystem.[23] Over recent years, hardly any plant or animal of non-Australian – especially of British – origin has escaped censure. The Australian salience of 'native = good', 'alien = bad', may reflect a special need to redress colonial Eurocentric preferences with chauvinist adoration of gum-trees and blowflies.[24]

With chauvinist exclusion goes worship of biological purity, to be saved from contamination by introduced aliens. But indigenous purity is neither possible nor desirable. The mixing of species, as of human races, is an unavoidable process and, in most contexts, its consequences are more desirable than otherwise. Nature and culture alike generally benefit from creative intermingling. Ex-colonial Jamaica, for example, readily domesticates what is alien. Since the seventeenth century, trees, grasses, crops and flowers brought in from the East Indies, Africa, North America and Europe have spread throughout the island. Do Jamaicans resent this riotous medley for displacing native flora? Quite the contrary; they rejoice in it as intrinsically Jamaican. They celebrate the commingled fragments of manifold ecologies enhanced by exotica from every land.[25]

The word 'imperial' today connotes the lacerations of alien imposition, the brutal harshness of indigenous extirpation, the callous uprooting or dismissive denial of autochthonous environments. At least some of the earlier meanings of 'empire' were more friendly. Emerson's idea of the 'imperial self' is grounded in ideals of sturdy self-awareness, taking possession of the imperium of one's own consciousness.[26] In gaining greater awareness of our imperial antecedents, we do better to embrace them within, rather than exile them from, our normative ecologies.

Notes

1. See Williams, Chapter 11, above.
2. See Thomas R. Dunlap, Chapter 5, above.
3. See Stephen J. Pyne, Chapter 1, above.
4. David Watts, *The West Indies: Patterns of Development, Culture and Environmental Change since 1492* (Cambridge: 1987); Jean Besson and Janet Momsen (eds), *Land and Development in the Caribbean* (London: 1987); David Lowenthal, 'Population Contrasts in the Guianas', *Geographical Review*, 50 (1960), pp. 41–58; B. Havard Duclos, *MABES: Pilot Economic Project for Grouping the Islands around St. Martin* (Caribbean Commission, Port-of-Spain, Trinidad: 1956). The fourth imperial power was the United States.
5. See Melville, Chapter 12, above.
6. Appointed 'Protector of the Indians' by Spain, Las Casas's eye-witness account (1519) of the atrocities that had decimated Hispaniola induced the Pope and the Holy Roman Emperor to ban American Indian slavery. The African slave trade was legitimated to protect the Indians: see Lawis Hanke, *Aristotle and the American Indians: A Study in Race Prejudice in the Modern World* (Chicago: 1959); Carl O. Sauer, *The Early Spanish Main* (Berkeley: 1966).
7. For a similar argument relating to Australia, see Libby Robin, Chapter 4, above.
8. See Griffiths, Introduction, above.

9. Bain Attwood, 'Mabo, Australia and the End of History', in *In the Age of Mabo: History, Aborigines and Australia* (St Leonards, NSW: 1996), pp. 100–16.
10. Martyn J. Bowden, 'The Great American Desert in the American Mind: The Historiography of a Geographical Notion', in David Lowenthal and Martyn J. Bowden (eds), *Geographies of the Mind* (New York: 1975), pp. 119–47 [American Geographical Society].
11. 'The Buffalo Commons Debate', *Focus* [American Geographical Society], 43/4 (1993), pp. 16–27.
12. T. H. Huxley, *Man's Place in Nature* (1863; Ann Arbor: 1959), p. 71.
13. O. H. K. Spate, 'Environmentalism', *International Encyclopedia of the Social Sciences* (New York: 1968), vol. 5, pp. 93–7.
14. George Perkins Marsh, *Man and Nature; or, Physical Geography as Modified by Human Action* (1864), ed. David Lowenthal (Cambridge, MA: 1965).
15. Clarence J. Glacken, *Traces on the Rhodian Shore: Nature and Culture in Western Thought from Ancient Times to the End of the Eighteenth Century* (Berkeley: 1967).
16. Jeremy Cohen, *Be Fruitful and Increase, Fill the Earth and Master It: The Ancient and Medieval Career of a Biblical Text* (Ithaca: 1989); Peter N. Carroll, *Puritanism and the American Wilderness* (New York: 1969).
17. David Lowenthal, 'Awareness of Human Impacts: Changing Attitudes and Emphases', in B. L. Turner, II, *et al.* (eds), *The Earth as Transformed by Human Action* (New York: 1990), pp. 121–35.
18. Carruthers, Chapter 8, above.
19. Alston Chase, *Playing God in Yellowstone* (San Diego: 1986); Terence Young, 'Virtue and Irony at Cades Cove', in *The Landscape of Theme Parks* (Washington, DC: forthcoming).
20. Kirkpatrick Sale, *The Conquest of Paradise: Christopher Columbus and the Columbian Legacy* (New York: 1990), pp. 368–9; Joan D. Laxson, 'How "we" See "them": Tourism and Native Americans', *Annals of Tourism Research*, 18 (1991), pp. 365–91; David Maybury-Lewis, *Millennium: Tribal Wisdom and the Modern World* (New York: 1992).
21. Jeffrey A. McNeely and William S. Keeton, 'The Interaction between Biological and Cultural Diversity', in Bernd von Droste, Harald Plachter, and Mechtild Rössler (eds), *Cultural Landscapes of Universal Value* (Jena: 1995), pp. 25–37.
22. David Lowenthal, *The Heritage Crusade and the Spoils of History* (London and Melbourne: 1997), pp. 81, 150–1, 194–8.
23. See Rolls and Flannery, Chapters 2 and 3, above.
24. Robin Boyd, *The Great Great Australian Dream* (Rushcutters Bay, NSW: 1972).
25. Derek Walcott, *The Antilles: Fragments of Epic Memory* (London: 1993).
26. Quentin Anderson, *The Imperial Self* (New York: 1971), p. 47.

Select Bibliography

Adas, Michael, *Machines as the Measure of Men: Science, Technology and Ideologies of Western Dominance* (Ithaca: 1989).

Allen, David Elliston, *The Naturalist in Britain: A Social History* (London: 1976).

Anderson, David and Richard Grove (eds), *Conservation in Africa: People, Policies and Practice* (Cambridge: 1987).

Anderson, Quentin, *The Imperial Self* (New York: 1971).

Arnold, David (ed.), *Imperial Medicine and Indigenous Societies* (Manchester: 1988).

—— and Ramachandra Guha (eds), *Nature, Culture, Imperialism: Essays in Indian Environmental History* (Delhi: 1995).

Attwood Bain (ed.), *In the Age of Mabo: History, Aborigines and Australia* (St Leonards: 1996).

Barakan, Elazar, *The Retreat of Scientific Racism: Changing Concepts of Race in Britain and the United States between the Wars* (Cambridge: 1992).

Barr, Neil and John Cary, *Greening a Brown Land: The Search for Sustainable Land Use* (Melbourne: 1992).

Basalla, George, 'The Spread of Western Science', *Science*, 156 (1967), pp. 611–22.

Beinart, William, 'Soil Erosion, Conservationism and Ideas about Development: A Southern African Exploration, 1900–1960', *Journal of Southern African Studies*, 11/1 (1984), pp. 52–83.

—— 'Empire, Hunting and Ecological Change in Southern and Central Africa,' *Past and Present*, 128 (1990), pp. 162–86.

—— 'La Nuit du chacal: moutons, pâturages et prédateurs en Afrique du Sud de 1900 à 1930', *Revue Française d'Histoire d'Outre-mer*, 80/298 (1993), pp. 105–29.

—— 'Environmental Destruction in Southern Africa: Soil Erosion, Animals and Pastures over the Longer Term', in Thackwray Driver and Graham Chapman (eds), *Timescales and Environmental Change* (London: 1996).

—— and Peter Coates, *Environment and History: The Taming of Nature in the USA and South Africa* (London: 1995).

Besson, Jean and Janet Momsen (eds), *Land and Development in the Caribbean* (London: 1987).

Birckhead, Jim, Terry de Lacy and Laurajane Smith (eds), *Aboriginal Involvement in Parks and Protected Areas* (Canberra: 1992).

Bolton, Geoffrey, *Spoils and Spoilers: Australians Make their Environment, 1788–1980* (Sydney: 1981).

Bonyhady, Tim and Tom Griffiths (eds), *Prehistory to Politics: John Mulvaney, the Humanities and the Public Intellectual* (Melbourne: 1996).

Bowden, Martyn J., 'The Great American Desert in the American Mind: The Historiography of a Geographical Notion', in David Lowenthal and Martyn J. Bowden (eds), *Geographies of the Mind* (New York: 1975), pp. 119–47.

Bowler, Peter J., *The Fontana History of the Environmental Sciences* (London: 1992).

Bozzoli, B. (ed.), *Town and Countryside in the Transvaal* (Johannesburg: 1983).

Brockway, Lucile H., *Science and Colonial Expansion: The Role of the British Royal Botanical Gardens* (New York: 1979).

Carruthers, Jane, 'Creating a National Park, 1910 to 1926', *Journal of Southern African Studies*, 15/2 (1989), pp. 188–216.

—— 'The Dongola Wild Life Sanctuary: "Psychological Blunder, Economic Folly and Political Monstrosity" or "More Valuable than Rubies and Gold"?', *Kleio*, 24 (1992), pp. 82–100.

—— 'Dissecting the Myth: Paul Kruger and the Kruger National Park', *Journal of Southern African Studies*, 20/2 (1994), pp. 263–83.

—— *The Kruger National Park: A Social and Political History* (Pietermaritzburg: 1995).

Cathcart, Michael, 'The Geography of Silence', *RePublica*, 3 (1995), pp. 178–88.

Caughley, Graeme, *The Deer Wars* (Auckland: 1983).

Chase, Alston, *Playing God in Yellowstone: The Destruction of America's First National Park* (San Diego: 1986).

Clark, Andrew Hill, *The Invasion of New Zealand by People, Plants and Animals* (New Brunswick: 1949).

Clark, Grahame, *Space, Time and Man: A Prehistorian's View* (Cambridge: 1994).

Cochrane, P., *Industrialization and Dependence* (St Lucia: 1980).

Cock, Jacklyn and Eddie Koch (eds), *Going Green: People, Politics and the Environment in South Africa* (Cape Town: 1991).

Collier, George A., Renato I. Rosaldo and John D. Wirth, *The Inca and Aztec States, 1400–1800: Anthropology and History* (New York: 1982).

Cook, S. B., *Imperial Affinities: Nineteenth Century Analogies and Exchanges between India and Ireland* (New Delhi: 1993).

Cowen, M. and R. Shenton, *Doctrines of Development* (London: 1996).

Cox, Thomas R., *Mills and Markets: A History of the Pacific Coast Lumber Industry to 1900* (Seattle: 1974).

Cranefield, Paul F., *Science and Empire: East Coast Fever in Rhodesia and the Transvaal* (Cambridge: 1991).

Cronon, William, *Changes in the Land: Indians, Colonists and the Ecology of New England* (New York: 1983).

—— *Nature's Metropolis: Chicago and the Great West* (New York: 1991).

—— George Miles and Jay Gitlin (eds), *Under an Open Sky: Rethinking America's Western Past* (New York: 1992).

Crosby, Alfred W., *The Columbian Exchange: Biological and Cultural Consequences of 1492* (Westport: 1972).

—— *Ecological Imperialism: The Biological Expansion of Europe 900–1900* (New York and Cambridge: 1986).

—— *Germs, Seeds and Animals: Studies in Ecological History* (New York: 1994).

—— 'The Past and Present of Environmental History', *American Historical Review* (1995), pp. 1177–89.

Crowcroft, Peter, *Elton's Ecologists* (Chicago: 1991).

Currie, George and John Graham, *The Origins of CSIRO: Science and the Commonwealth Government 1901–1926* (Melbourne: 1966).

Curtin, Philip D., *Cross-Cultural Trade in World History* (Cambridge: 1990).

—— *The Rise and Fall of the Plantation Complex: Essays in Atlantic History* (Cambridge: 1990).

Cutliffe, S. and Robert Post (eds), *In Context: History and the History of Technology* (London: 1989).

Dargavel, John and Richard Tucker (eds), *Changing Pacific Forests: Historical Perspectives on the Forest Economy of the Pacific Basin* (Durham, NC: 1992).

Darian-Smith, Kate, Liz Gunner and Sarah Nuttall (eds), *Text, Theory, Space: Land, Literature and History in South Africa and Australia* (London, 1996).

Davison, Graeme, *The Unforgiving Minute: How Australia Learned to Tell the Time* (Melbourne: 1993).

Dean, Warren, *With Broadax and Firebrand: The Destruction of the Brazilian Atlantic Forest* (Berkeley: 1995).

Degler, Carl, *In Search of Human Nature: The Decline and Revival of Darwinism in American Social Thought* (New York: 1991).

Denoon, Donald, *Settler Capitalism: The Dynamics of Dependent Development in the Southern Hemisphere* (Oxford: 1983).

—— 'The Isolation of Australian History', *Historical Studies*, 22/87 (1986), pp. 252–60.

—— 'Settler Capitalism Unsettled', *New Zealand Journal of History*, 29/2 (1995), pp. 129–41.

Dixon, C. and M. Heffernan (eds), *Colonialism and Development in the Contemporary World* (London: 1991).

Dixon, R., *Writing the Colonial Adventure: Race, Gender and Nation in Anglo-Australian Popular Fiction, 1875–1914* (Cambridge: 1995).

Dovers, Stephen (ed), *Australian Environmental History: Essays and Cases* (Melbourne: 1994).

Driver, Thackwray and Graham Chapman (eds), *Timescales and Environmental Change* (London: 1996).

Droste, Bernd von, Harald Plachter and Mechtild Rössler (eds), *Cultural Landscapes of Universal Value* (Jena: 1995).

Dunlap, Thomas R., *Saving America's Wildlife: Ecology and the American Mind* (Princeton, NJ: 1988).

—— 'Australian Nature, European Culture: Anglo Settlers in Australia', *Environmental History Review*, 17/1 (1993), pp. 24–48.

Dyster, Barrie and David Meredith, *Australia in the International Economy* (Melbourne: 1990).

Ellis, Stephen, 'Of Elephants and Men: Politics and Nature Conservation in South Africa', *Journal of Southern African Studies*, 20/1 (1993), pp. 53–69.

Everhart, W. C., *The National Park Service*, 2nd edn (Boulder: 1983).

Flader, Susan L. (ed.), *The Great Lakes Forest: An Environmental and Social History* (Minneapolis: 1983), pp. 253–73.

Flannery, Timothy Fritjof, *The Future Eaters* (Sydney: 1994; reprinted 1995).

Freemuth, John C., *Islands under Siege: National Parks and the Politics of External Threats* (Lawrence: 1991).

Gadgil, Madhav and Ramachandra Guha, 'State Forestry and Social Conflict in British India', *Past and Present*, 123 (1989), pp. 141–77.

—— *This Fissured Land: An Ecological History of India* (Oxford: 1992).

—— *Ecology and Equity: The Use and Abuse of Nature in Contemporary India* (London: 1995).

Galbreath, Ross, *Working for Wildlife* (Wellington: 1993).

Gale, Fay and Graham H. Lawton (eds), *Settlement and Encounter: Geographical Studies Presented to Sir Grenfell Price* (Melbourne: 1969).

Glacken, Clarence J., *Traces on the Rhodian Shore: Nature and Culture in Western Thought from Ancient Times to the End of the Eighteenth Century* (Berkeley: 1967).

Griffiths, Tom, *Secrets of the Forest* (Sydney: 1992).

—— *Hunters and Collectors: The Antiquarian Imagination in Australia* (Melbourne and Cambridge: 1996).

—— 'In Search of Australian Antiquity', in Tim Bonyhady and Tom Griffiths (eds), *Prehistory to Politics: John Mulvaney, the Humanities and the Public Intellectual* (Melbourne: 1996), pp. 42–62.

Grove, Richard, 'Early Themes in African Conservation: The Cape in the Nineteenth Century', in David Anderson and Richard Grove (eds), *Conservation in Africa* (Cambridge: 1987), pp. 21–39.

—— 'Scottish Missionaries, Evangelical Discourses and the Origins of Conservation Thinking in Southern Africa 1820–1900', *Journal of Southern African Studies*, 15/2 (1989), pp. 163–87.

—— *Green Imperialism: Colonial Expansion, Tropical Island Edens and the Origins of Environmentalism, 1600–1860* (Cambridge: 1995).

Hays, Samuel P., *Conservation and the Gospel of Efficiency: The Progressive Conservation Movement 1890–1920* (Cambridge, MA: 1959).

—— *Beauty, Health and Permanence: Environmental Politics in the United States 1955–1985* (Cambridge, MA: 1987)

Headrick, Daniel R., *The Tools of Empire: Technology and European Imperialism in the Nineteenth Century* (Oxford and New York: 1981).

—— *The Tentacles of Progress: Technology Transfer in the Age of Imperialism, 1850–1940* (Oxford and New York: 1988).

—— *The Hidden Weapon: Telecommunications and International Politics, 1851–1945* (Oxford and New York: 1991).

Heathcote, R. L. (ed.), *The Australian Experience: Essays in Australian Land Settlement and Resource Management* (Melbourne: 1988).

—— *Australia* (1975; Harlow, Essex: 1994).

Hill, Kevin A., 'Conflicts over Development and Environmental Values: The International Ivory Trade in Zimbabwe's Historical Context', *Environment and History*, 1/3 (1995), pp. 335–49.

Hirst, J. B., 'Keeping Colonial History Colonial: The Hartz Thesis Revisited', *Historical Studies*, 21/82 (1984), pp. 85–104.

Home, R. W., *Australian Science in the Making* (Cambridge: 1988).

Inkster, Ian, 'Scientific Enterprise and the Colonial "Model": Observations on Australian Experience in Historical Context', *Social Studies of Science*, 15 (1985), pp. 677–704.

Ishwaran, N., 'Biodiversity, Protected Areas and Sustainable Development', *Nature and Resources*, 28/1 (1992), pp. 18–25.

Jacks, G. V. and R. O. Whyte, *The Rape of the Earth: A World Survey of Soil Erosion* (London: 1939).

Journal of Southern African Studies, 15/2 (1989): special issue on conservation in southern Africa, with an introduction by William Beinart, 'The Politics of Colonial Conservation'.

Khan, Farieda, 'Rewriting South Africa's Conservation History: The Role of the Native Farmers Association', *Journal of Southern African Studies*, 20/4 (1994), pp. 499–516.

Kjekshus, Helge, *Ecology, Control and Economic Development in East African History* (London: 1977).

Kluger, J. R., *Turning on Water with a Shovel: The Career of Elwood Mead* (Albuquerque: 1992).

Latz, Peter, *Bushfires and Bushtucker: Aboriginal Plant Use in Central Australia* (Alice Springs: 1995).

Leeper, Geoffrey W. (ed.), *The Australian Environment* (Canberra: 1970).

Leopold, Aldo, *A Sand County Almanac* (1948; New York: 1966).

Limerick, Patricia Nelson, *The Legacy of Conquest: The Unbroken Past of the American West* (New York: 1987).

Lines, William, *Taming the Great South Land: A History of the Conquest of Nature in Australia* (Sydney: 1991).

Lowenthal, David, *George Perkins Marsh: Versatile Vermonter* (New York: 1958).

—— 'Population Contrasts in the Guianas', *Geographical Review*, 50 (1960), pp. 41–58.

—— 'Awareness of Human Impacts: Changing Attitudes and Emphases', in B. L. Turner, II, *et al.* (eds), *The Earth as Transformed by Human Action* (New York: 1990), pp. 121–35.

—— *The Heritage Crusade and the Spoils of History* (London: 1997).

—— and Martyn J. Bowden (eds), *Geographies of the Mind* (New York: 1975).

Mabey, Richard, *The Common Ground* (London: 1980).

McIntosh, Robert M., *The Background of Ecology* (New York: 1985).

MacKenzie, John M. (ed.), *Imperialism and Popular Culture* (Manchester: 1986).

—— *The Empire of Nature: Hunting, Conservation and British Imperialism* (Manchester: 1988).

—— *Imperialism and the Natural World* (Manchester: 1990).

—— *Orientalism: History, Theory and the Arts* (Manchester: 1995).

MacLeod, Roy M., 'Scientific Advice for British India', *Modern Asian Studies*, 9/3 (1975), pp. 343–84.

—— 'On Visiting the "Moving Metropolis"', *Historical Records of Australian Science*, 5/1 (1982), pp. 1–16.

—— *The Commonwealth of Science* (Melbourne: 1988).

—— and M. Lewis (eds), *Disease, Medicine and Empire* (London: 1988).

Mann, Michael, 'Ecological Change in North India: Deforestation and Agrarian Distress in the Ganga-Jamna Doab 1800–1850', *Environment and History*, 1/2 (1995), pp. 201–20.

Mannion, A. M., *Global Environmental Change: A Natural and Cultural Environmental History* (London: 1991).

Marks, Robert B., 'Commercialization without Capitalism: Processes of Environmental Change in South China, 1550–1850', *Environmental History*, 1/1 (1996) pp. 56–82.

Marks, Stuart A., *The Imperial Lion: Human Dimensions of Wildlife Management in Central Africa* (Boulder: 1984).

Markus, Andrew, *Australian Race Relations* (St Leonards: 1994).

Marsh, George Perkins, *Man and Nature*, ed. David Lowenthal (Cambridge, MA: 1965).

Maybury-Lewis, David, *Millennium: Tribal Wisdom and the Modern World* (New York: 1992).

Meinig, Donald W., 'A Macrogeography of Western Imperialism: Some Morphologies of Moving Frontiers of Political Control', in Fay G. Gale and Graham H. Lawton (eds.), *Settlement and Encounter: Geographical Studies Presented to Sir Grenfell Price* (Melbourne: 1969), pp. 213–40.

—— *On the Margins of the Good Earth: The South Australian Wheat Frontier 1869–1884* (1962; Adelaide: 1970).

Melville, Elinor G. K., *A Plague of Sheep: Environmental Consequences of the Conquest of Mexico* (New York and Cambridge: 1994).

Merchant, Carolyn, 'The Theoretical Structure of Ecological Revolutions', *Environmental Review*, 11/4 (1987), pp. 265–74.

Mitman, Gregg, *The State of Nature: Ecology, Community and American Social Thought: 1900–1950* (Chicago: 1992).

Nash, Roderick F., The American Invention of National Parks', *American Quarterly*, 23/3 (1970), pp. 726–35.

—— *The American Environment: Readings in the History of Conservation*, 2nd edn (Reading, MA: 1976).

—— *Wilderness and the American Mind* (1969; New Haven: 1982).

Nelson, J. G., R. D. Needham and D. L. Mann (eds), *International Experience with National Parks and Related Reserves* (Waterloo, Ont.: 1978).

O'Brien, Patrick, 'European Economic Development: The Contribution of the Periphery', *The Economic History Review*, 2nd Series, 35/1 (1982).

Packard, Randall M., *White Plague, Black Labour: Tuberculosis and the Political Economy of Health and Disease in South Africa* (Pietermaritzburg: 1990).

Park, Geoff, *Ñga Uruora (The Groves of Life): Ecology and History in a New Zealand Landscape* (Wellington: 1995).

Patch, Robert, 'Imperial Politics and Local Economy in Colonial Central America, 1670–1770', *Past and Present*, 143 (1996) pp. 77–107.

Pauly, Philip J., 'The Beauty and Menace of the Japanese Cherry Trees: Conflicting Visions of American Ecological Independence', *Isis*, 87 (1996), pp. 51–73.

Pepper, David, *The Roots of Modern Environmentalism* (London: 1984).

Perdue, Peter C., *Exhausting the Earth: State and Peasant in Hunan, 1500–1850* (Cambridge, MA: 1987).

Pisani, Donald J., *To Reclaim a Divided West: Water, Law, and Public Policy, 1848–1902* (Albuquerque: 1992).

Player, I. (ed.), *Voices of the Wilderness* (Johannesburg: 1979).

Ponting, Clive, *A Green History of the World* (Harmondsworth: 1991).

Powell, J. M., *The Public Lands of Australia Felix* (Melbourne: 1970).

—— 'Conservation and Resource Management in Australia 1788–1860', in J. M. Powell and Michael Williams (eds), *Australian Space, Australian Time* (Melbourne: 1975), pp. 18–60.

—— *Environmental Management in Australia 1788–1914* (Melbourne: 1976).

—— 'Archibald Grenfell Price (1892–1977)', in T. W. Freeman (ed.), *Geographers: Bibliographical Studies*, 6 (1982), pp. 87–92.

—— 'Elwood Mead and California's State Colonies: An Episode in Australasian–American Contacts, 1915–31', *Journal of the Royal Australian Historical Society*, 67 (1982), pp. 328–53.

—— *Mirrors of the New World* (Canberra: 1978).

—— *An Historical Geography of Modern Australia: The Restive Fringe* (Cambridge: 1988).

—— *Watering the Garden State: Land, Water and Community in Victoria 1834–1988* (Sydney: 1989).

—— *Plains of Promise, Rivers of Destiny* (Brisbane: 1991).

—— *Griffith Taylor and 'Australia Unlimited'* (St Lucia: 1993).

—— *'MDB': The Emergence of Bioregionalism in the Murray–Darling Basin* (Canberra: 1993).

—— 'Historical Geography and Environmental History: An Australian Interface', *Journal of Historical Geography*, 22/3 (1996), pp. 253–73.

Price, A. Grenfell, *White Settlers and Native Peoples* (Melbourne: 1950).

—— *The Western Invasions of the Pacific and its Continents: A Survey of Moving Frontiers and Changing Landscapes, 1513–1958* (Oxford: 1963).

—— *The Importance of Disease in History* (Adelaide: 1964).

Pyne Stephen J., *The Ice: A Journey to Antarctica* (Iowa City: 1986).

—— *Burning Bush: A Fire History of Australia* (New York: 1991).

—— *Vestal Fire: An Environmental History, Told through Fire, of Europe and of Europe's Encounter with the World* (Seattle: forthcoming).

Rangarajan, Mahesh, 'Environmental Histories of South Asia', *Environment and History*, 2/2 (1996), pp. 129–43.

Ratcliffe, Francis, *Flying Fox and Drifting Sand: The Adventures of a Biologist in Australia* (London: 1938).

Reid, J. R. W., J. A. Kerle and S. R. Morton (eds), *Uluru Fauna* (Canberra: 1993).

Reingold, Nathan and Marc Rothenberg (eds), *Scientific Colonialism* (Washington, DC: 1987).

Reynolds, Henry, *Frontier: Aborigines, Settlers and Land* (Sydney: 1987).

Ritvo, Harriet, *The Animal Estate: The English and Other Creatures in the Victorian Age* (1986; Harmondsworth: 1990).

Roberts, Andrew, *The Colonial Moment in Africa* (Cambridge: 1990).

Robin, Libby, 'Of Desert and Watershed: The Rise of Ecological Consciousness in Victoria, Australia', in Michael Shortland (ed.), *Science and Nature* (Oxford: 1993), pp. 115–49.

—— 'Nature Conservation as a National Concern: The Role of the Australian Academy of Science', *Historical Records of Australian Science*, 10/1 (1994), pp. 1–24.

—— 'Radical Ecology and Conservation Science: An Australian Perspective', *Environment and History* (1997, forthcoming).

Robinson, Robbie (ed.), *African Heritage 2000: The Future of Protected Areas in Africa* (Pretoria: 1995).

Roche, Michael, *Land and Water: Water and Soil Conservation and Central Government in New Zealand 1941–1988* (Wellington: 1994).

Rolls, Eric C., *They All Ran Wild* (1969: 1984).

—— *A Million Wild Acres* (Melbourne: 1981).

—— *Sojourners* (St Lucia: 1992).

—— *Citizens* (St Lucia: 1996).

Romano, Ruggiero, *La Crise du XVIIe siècle* (Geneva: 1992).

Runte, Alfred, *National Parks: The American Experience* (Lincoln: 1979).

Said, Edward, *Orientalism* (Harmondsworth: 1985).

Sale, Kirkpatrick, *The Conquest of Paradise: Christopher Columbus and the Columbian Legacy* (New York: 1990).

Salvucci, Richard J. (ed.), *Latin America and the World Economy: Dependency and Beyond* (London: 1996).

Sauer, Carl O., 'Destructive Exploitation in Modern Colonial Expansion', *Comptes Rendus du Congrès International de Géographie*, vol. 2 (Sect. 3c) (Amsterdam: 1938), pp. 494–99.

—— 'Theme of Plant and Animal Destruction in Economic History', *Journal of Farm Economics*, 20 (1938), pp. 765–75.

Schama, Simon, *Landscape and Memory* (London: 1995).

Schedvin, C. B., *Shaping Science and Industry: A History of Australia's CSIR 1926–1949* (Sydney: 1987).

Schmitt, P. J., *Back to Nature: The Arcadian Myth in Urban America* (New York: 1969).

Shortland, Michael (ed.), *Science and Nature: Essays in the History of the Environmental Sciences* (Oxford: 1993).

Simmons, I. G., *Environmental History: A Concise Introduction* (Oxford: 1993).

Sivaramakrishnan, K., 'Colonialism and Forestry in India: Imagining the Past and Present Politics', *Comparative Studies in History and Society*, 37 (1995), pp. 3–40.

Smith, A., *Pastoralism in Africa: Origins and Development Ecology* (Johannesburg: 1992).

Smout, T. C., 'The Highlands and the Roots of Green Consciousness, 1750–1990', the Raleigh Lecture on History, *Proceedings of the British Academy*, 76 (1991), pp. 237–63.

Stepan, Nancy, *The Idea of Race in Science: Great Britain 1800–1960* (London: 1982).

Stern, Steve J., *Peru's Indian Peoples and the Challenge of the Spanish Conquest* (Madison: 1993).

Thomas, Keith, *Man and the Natural World: Changing Attitudes in England, 1500–1800* (London: 1983).

Thompson, Harry V. and Carolyn King (eds), *The European Rabbit* (Oxford: 1994).

Tobey, Ronald, *Saving the Prairies* (Berkeley: 1981).

Todd, Jan, *Colonial Technology: Science and the Transfer of Innovation to Australia* (Cambridge: 1995).

Totman, Conrad, *The Green Archipelago: Forestry in Preindustrial Japan* (Berkeley: 1989).

Trainor, Luke, *British Imperialism and Australian Nationalism* (Cambridge: 1994).

Tucker, Richard P. and John F. Richards (eds), *Global Deforestation in the Nineteenth-Century World Economy* (Durham, NC: 1988).

Turner, B. L., II, *et al.* (eds), *The Earth as Transformed by Human Action: Global and Regional Changes in the Biosphere over the Last 300 Years* (New York: 1990).

Vail, Leroy, 'Ecology and History: The Example of Eastern Zambia', *Journal of Southern African Studies*, 3 (1977), pp. 129–55.

Vaughan, M. *Curing their Ills* (Cambridge: 1991).

Watts, David, *The West Indies: Patterns of Development, Culture and Environmental Change since 1492* (Cambridge: 1987).

West, Patrick C. and Steven R. Brechin (eds), *Resident Peoples and National Parks: Social Dilemmas and Strategies in International Conservation* (Tucson: 1991).

Western, David and R. Michael Wright, *Natural Connections: Perspectives in Community-Based Conservation* (Washington, DC: 1994).

Whitcombe, Elizabeth, *Agrarian Conditions in North India* (Berkeley: 1972).

White, M. E., *After the Greening: The Browning of Australia* (Sydney: 1994).

White, Richard, *'It's your Misfortune and None of my Own': A New History of the American West* (Norman and London: 1991).

Wilkinson, Lise, *Animals and Disease: An Introduction to the History of Comparative Medicine* (Cambridge: 1992).

Williams, Michael, *The Making of the South Australian Landscape* (London: 1974).

—— 'Thinking about the Forest: A Comparative View from Three Continents', in Susan L. Flader (ed.), *The Great Lakes Forest: An Environmental and Social History* (Minneapolis: 1983), pp. 253–73.

—— 'The Clearing of the Woods', in R. L. Heathcote (ed.), *The Australian Experience: Essays in Australian Land Settlement and Resource Management* (Melbourne: 1988), pp. 115–26.

—— *The Americans and their Forests: A Historical Geography* (New York: 1989).

—— 'Environmental History and Historical Geography', *Journal of Historical Geography*, 20/1 (1994), pp. 3–21.

—— 'European Expansion and Land Cover Transformation', in Ian Douglas, Richard Huggett and Mike Robinson (eds), *Companion Encyclopedia of Geography* (London: 1996), pp. 182–205.

Wolf, Eric, *Europe and the People without History* (Berkeley: 1982).

Worboys, Michael, 'Science and British and Colonial Imperialism 1895–1940', PhD thesis (University of Sussex: 1979).

—— 'The Discovery of Colonial Malnutrition', in D. Arnold (ed.), *Imperial Medicine and Indigenous Societies* (Manchester: 1988), pp. 208–25.

Worster, Donald, *Dust Bowl: The Southern Plains in the 1930s* (New York and Oxford: 1979).

—— *Rivers of Empire: Water, Aridity and the Growth of the American West* (New York: 1985).

—— *The Ends of the Earth: Perspectives on Modern Environmental History* (Cambridge: 1988).

—— *Under Western Skies: Nature and History in the American West* (New York and Oxford: 1992).

—— *Nature's Economy: A History of Ecological Ideas* (1977; Cambridge: 1994).

Notes on Contributors

William Beinart is Professor in Race Relations in Social Studies and History at the University of Oxford and co-author (with Peter Coates) of *Environment and History: The Taming of Nature in the USA and South Africa* (1995).

Jane Carruthers is a historian at the University of South Africa and author of *The Kruger National Park: A Social and Political History* (1995).

Thomas R. Dunlap is Professor of History at Texas A & M University. His books include *Saving America's Wildlife: Ecology and the American Mind* (1988).

Timothy F. Flannery is Senior Research Scientist in Mammalogy at the Australian Museum, Sydney, and author of *The Future Eaters* (1994).

Tom Griffiths is Fellow, History Program, in the Research School of Social Sciences, Australian National University, Canberra, and former deputy head of the Sir Robert Menzies Centre for Australian Studies, University of London. He is author of *Hunters and Collectors* (1996) and *Secrets of the Forest* (1992).

Richard Grove is Senior Fellow in Economic History in the Research School of Social Sciences, Australian National University, Canberra, and author of *Green Imperialism* (1995).

Brigid Hains lectures in history at Monash University, Victoria, Australia.

David Lowenthal is Emeritus Professor in Geography at University College, London. His books include *The Past is a Foreign Country* and *The Heritage Crusade and the Spoils of History* (1997).

John M. MacKenzie is Professor of History at Lancaster University, England. He is author of *The Empire of Nature* (1988) and editor of *Imperialism and the Natural World* (1990).

Elinor G. K. Melville is Professor of History at York University, Canada, and author of *A Plague of Sheep: Environmental Consequences of the Conquest of Mexico* (1994).

Shaun Milton is a historian at the Institute of Commonwealth Studies, University of London.

Joseph M. Powell is Professor of Geography in the Department of Geography and Environmental Science at Monash University, Australia. He is author of *An Historical Geography of Modern Australia: The Restive Fringe* (1988) and *Mirrors of the New World* (1978), amongst others.

Stephen J. Pyne is Professor of History at Arizona State University. His books include *Burning Bush: A Fire History of Australia* (1991) and *Vestal Fire: An Environmental History, Told Through Fire, of Europe and of Europe's Encounter with the World* (forthcoming).

Libby Robin is an Australian Research Council Postdoctoral Fellow at the Humanities Research Centre, Australian National University, Canberra. She is author of *Building a Forest Conscience*.

Eric Rolls is a farmer, poet and historian, now living at North Haven, New South Wales. His historical books include *They All Ran Wild* (1969), *A Million Wild Acres* (1981), *Sojourners* (1992) and *Citizens* (1996).

Michael Williams is a Professor in the School of Geography, University of Oxford. He is author of *Americans and their Forests* (1989) and *The Making of the South Australian Landscape* (1974).

Index